Logic
DeMYSTiFieD®

DeMYSTiFieD® Series

Accounting Demystified
Advanced Calculus Demystified
Advanced Physics Demystified
Advanced Statistics Demystified
Algebra Demystified
Alternative Energy Demystified
Anatomy Demystified
asp.net 2.0 Demystified
Astronomy Demystified
Audio Demystified
Biology Demystified
Biotechnology Demystified
Business Calculus Demystified
Business Math Demystified
Business Statistics Demystified
C++ Demystified
Calculus Demystified
Chemistry Demystified
Circuit Analysis Demystified
College Algebra Demystified
Corporate Finance Demystified
Databases Demystified
Data Structures Demystified
Differential Equations Demystified
Digital Electronics Demystified
Earth Science Demystified
Electricity Demystified
Electronics Demystified
Engineering Statistics Demystified
Environmental Science Demystified
Everyday Math Demystified
Fertility Demystified
Financial Planning Demystified
Forensics Demystified
French Demystified
Genetics Demystified
Geometry Demystified
German Demystified
Home Networking Demystified
Investing Demystified
Italian Demystified
Java Demystified
JavaScript Demystified
Lean Six Sigma Demystified
Linear Algebra Demystified
Logic Demystified
Macroeconomics Demystified
Management Accounting Demystified
Math Proofs Demystified
Math Word Problems Demystified
MATLAB® Demystified
Medical Billing and Coding Demystified
Medical Terminology Demystified
Meteorology Demystified
Microbiology Demystified
Microeconomics Demystified
Nanotechnology Demystified
Nurse Management Demystified
OOP Demystified
Options Demystified
Organic Chemistry Demystified
Personal Computing Demystified
Pharmacology Demystified
Physics Demystified
Physiology Demystified
Pre-Algebra Demystified
Precalculus Demystified
Probability Demystified
Project Management Demystified
Psychology Demystified
Quality Management Demystified
Quantum Mechanics Demystified
Real Estate Math Demystified
Relativity Demystified
Robotics Demystified
Sales Management Demystified
Signals and Systems Demystified
Six Sigma Demystified
Spanish Demystified
sql Demystified
Statics and Dynamics Demystified
Statistics Demystified
Technical Analysis Demystified
Technical Math Demystified
Trigonometry Demystified
uml Demystified
Visual Basic 2005 Demystified
Visual C# 2005 Demystified
XML Demystified

Logic DeMYSTiFieD®

Anthony Boutelle

Stan Gibilisco

New York Chicago San Francisco Lisbon London Madrid Mexico City
Milan New Delhi San Juan Seoul Singapore Sydney Toronto

The McGraw-Hill Companies

Cataloging-in-Publication Data is on file with the Library of Congress

Logic DeMYSTiFieD®

1 2 3 4 5 6 7 8 9 0 DOC/DOC 1 9 8 7 6 5 4 3 2 1 0

ISBN 978-0-07-170128-0
MHID 0-07-170128-1

Sponsoring Editor
Judy Bass

Acquisitions Coordinator
Michael Mulcahy

Editing Supervisor
David E. Fogarty

Project Manager
Tania Andrabi,
Glyph International

Copy Editor
Surendra Nath Shivam

Proofreader
Shivani Arora

Production Supervisor
Richard C. Ruzycka

Composition
Glyph International

Art Director, Cover
Jeff Weeks

Cover Illustration
Lance Lekander

About the Authors

Anthony Boutelle is a graduate of Macalester College in St. Paul, Minnesota, with a major in philosophy and a minor in physics. He currently works as a technical reviewer and editor for a computer firm in Minneapolis.

Stan Gibilisco is an electronics engineer, researcher, and mathematician who has authored several titles for the McGraw-Hill *Demystified* and *Know-It-All* series, along with dozens of other books on diverse technical topics.

Contents

Preface *xvii*

CHAPTER 1 **Arguments, Validity, and Truth Tables** **1**

What's Logic About? 2

Terminology 2

Logical Arguments 2

What's Logic Good For? 3

Logical Form 3

An Example 4

When Truth Doesn't Matter 4

When Meaning Doesn't Matter 4

Rules for Reasoning 5

Identity 5

Contradiction 6

True or False 6

Sentence Forms 6

Subject/Verb (SV) Statements 6

Subject/Verb/Object (SVO) Statements 7

Subject/Linking Verb/Complement (SLVC) Statements 7

Symbols and Operations 8

Logical Negation (NOT) 8

Logical Conjunction (AND) 9

Inclusive Logical Disjunction (OR) 9

Exclusive Logical Disjunction (XOR) 10

Logical Implication (IF/THEN) 11

Logical Equivalence (IFF) 12

Tables for Basic Operations 14
Table for Logical Negation 14
Table for Conjunction 14
Tables for Logical Disjunction 15
Table for Logical Implication 15
Table for Logical Equivalence 16
A Quick Proof 17
Precedence 18
Examples of Precedence 19
Proofs Using Truth Tables 20
Reversing the Order of a Conjunction 21
Grouping of Conjunctions 21
Reversing the Order of an Inclusive Disjunction 23
Grouping of Inclusive Disjunctions 23
Reversing the Order of an Implication 25
Ungrouping the Negation of a Conjunction 25
Ungrouping the Negation of an Inclusive Disjunction 26
Regrouping with Mixed Operations 27
Truth Tables Have Power! 27
Quiz 31

CHAPTER 2 **Propositional Logic** **33**
A Formal System 34
Propositional Formulas (PFs) 34
Is This Mathematics? 35
Contradiction 35
Sequents 36
Backing Up the Argument 37
Assumption (A) 37
Double Negation (DN) 38
Conjunction Introduction (∧I) 39
Conjunction Elimination (∧E) 39
Disjunction Introduction (∨I) 40
Disjunction Elimination (∨E) 40
Conditional Proof (CP) and Theorems 42
Biconditional Introduction (↔I) 44
Biconditional Elimination (↔E) 45
Substitution 45
Laws 48
Law of Contradiction 48
Law of Excluded Middle 48

Law of Double Negation 49
Commutative Laws 49
Associative Laws 50
Law of Implication Reversal 51
DeMorgan's Laws 53
Distributive Laws 54
Interderivability (⊣⊢), Deriving Laws, and Theorem Introduction (TI) 54
Quiz 57

CHAPTER 3 **Predicate Logic** **61**
Symbolizing Sentence Structure 62
Formal Predicates 62
Formal Subjects: Names 62
Formal Subjects: Variables 62
Predicate Sentence Formulas 63
More Predicate Sentences 64
Identity (=) 65
Building More Complex Formulas 65
Quantified Statements 66
Some versus All 66
The Existential Quantifier (∃) 67
The Universal Quantifier (∀) 68
The Universe 68
Multiple Quantifiers 70
Translating the Quantifiers 72
Laws of Quantifier Transformation 72
Logical Relationships: The Square of Opposition 73
Well-Formed Formulas 74
Properties of Two-Part Relations 75
Symmetry 75
Asymmetry 76
Antisymmetry 76
Nonsymmetry 76
Reflexivity 77
Irreflexivity 77
Nonreflexivity 77
Transitivity 77
Intratransitivity 78
Nontransitivity 78
Equivalence Relations 78

Predicate Proofs 78
Existential Introduction (∃I) 79
Existential Elimination (∃E) 79
Universal Introduction (∀I) 80
Universal Elimination (∀E) 81
Identity Introduction (=I) 81
Identity Elimination (=E) 82
Conventions for Predicate Proofs 82
Syllogisms 86
A Sample Syllogism 87
Allowed Propositions 87
Terms and Arguments 88
Conversion Laws 89
Classifying Syllogisms: Mood and Figure 89
Testing Syllogistic Arguments 90
Valid Syllogisms 93
Other Kinds of Syllogism 93
Quiz 95

CHAPTER 4 **A Boot Camp for Rigor** **99**
Definitions 100
Elementary Terms 100
Line Segment 100
Half-Open Line Segment 101
Open Line Segment 101
Length of Line Segment 102
Closed-Ended Ray 102
Open-Ended Ray 102
Point of Intersection 102
Collinear Points 103
Coplanar Points 103
Coincident Lines 103
Collinear Line Segments and Rays 103
Transversal 104
Parallel Lines 104
Parallel Line Segments 104
Parallel Rays 104
Angle 106
Measure of Angle 108
Straight Angle 108
Supplementary Angles 109
Right Angle 109

Complementary Angles 109
Perpendicular Lines, Line Segments, and Rays 109
Triangle 110
Similar and Congruent Triangles 112
Direct Similarity 112
Inverse Similarity 113
Direct Congruence 115
Inverse Congruence 117
Two Crucial Facts 118
Two More Crucial Facts 119
Axioms 120
The Two-Point Axiom 120
The Extension Axiom 120
The Right Angle Axiom 120
The Parallel Axiom 121
The Side-Side-Side (SSS) Axiom 121
The Side-Angle-Side (SAS) Axiom 122
The Angle-Side-Angle (ASA) Axiom 123
The Side-Angle-Angle (SAA) Axiom 123
Some Proofs at Last 125
Alternate Interior Angles 130
Quiz 142

CHAPTER 5 Fallacies, Paradoxes, and Revelations 145
The Probability Fallacy 146
Belief 146
Parallel Worlds, Fuzzy Worlds 146
We Must Observe 147
Weak and Flawed Reasoning 148
"Proof" by Example 148
Begging the Question 150
Hasty Generalization 151
Misuse of Context 151
Circumstance 151
Fallacies with Syllogisms 152
Fun with Silliness 154
Inductive Reasoning 155
Simple Paradoxes 158
A Wire around the Earth 158
Direct-Contradiction Paradox 160
Who Shaves Hap? 161

Arrow Paradox 161
The Frog and the Wall 162
A Geometry Trick 165
A "proof" that $-1 = 1$ 167
Wheel Paradox 169
Classical Paradoxes 170
Execution Paradox 170
A Two-Pronged Defense 171
Saloon Paradox 172
Barbershop Paradox 173
Shark Paradox 174
Russell and Gödel 175
Sets 175
Two Special Sets 175
The Paradox 176
Professor N's Machine 177
What Did You Prove? 178
Quiz 179

CHAPTER 6 **Strategies for Proofs** **181**
How Does a Theory Evolve? 182
Definitions 182
Elementary Terms 183
Axioms 184
Euclid's Postulates 184
A Denial 186
Propositions 187
Theorems 187
A Classical Theorem 188
Lemmas 189
Corollaries 189
Proofs, Truth, and Beauty 192
Techniques 194
Deductive Reasoning 194
What's the Universe? 195
Weak Theorems 196
Demonstrating a Weak Theorem 196
Strong Theorems 197
Demonstrating a Strong Theorem 198
Reductio ad Absurdum Revisited 199
Mathematical Induction 201

Cause, Effect, and Implication 205
Correlation and Causation 205
Complications 208
Quiz 212

CHAPTER 7 **Boolean Algebra** **215**
New Symbols for Old Operations 216
The NOT Operation (−) 216
The AND Operation (×) 216
The OR Operation (+) 217
Boolean Implication (⇒) 218
Boolean Equivalence (=) 219
Truth Tables, Boolean Style 220
Truth Table for Boolean Negation (NOT) 221
Truth Table for Boolean Multiplication (AND) 221
Truth Table for Boolean Addition (OR) 221
Truth Table for Boolean Implication (IF/THEN) 222
Truth Table for Boolean Equality (IFF) 222
Basic Boolean Laws 227
Precedence 227
Contradiction 228
Law of Double Negation 228
Commutative Laws 228
Associative Laws 229
Law of the Contrapositive 229
DeMorgan's Law for Products 230
DeMorgan's Law for Sums 230
Distributive Law 230
Quiz 236

CHAPTER 8 **The Logic of Sets** **239**
Set Fundamentals 240
To Belong, or Not to Belong 240
Listing the Elements 240
The Empty Set 241
Finite or Infinite? 241
Sets within Sets 242
Venn Diagrams 244
People and Numbers 244
Subsets 245

Proper Subsets 245
Congruent Sets 246
Disjoint Sets 247
Overlapping Sets 248
Set Intersection 253
Intersection of Two Congruent Sets 253
Intersection with the Null Set 253
Intersection of Two Disjoint Sets 253
Intersection of Two Overlapping Sets 254
Set Union 256
Union of Two Congruent Sets 257
Union with the Null Set 257
Union of Two Disjoint Sets 257
Union of Two Overlapping Sets 258
Quiz 263

CHAPTER 9 **The Logic of Machines** **265**
Numeration Systems 266
Decimal 267
Binary 267
Octal 268
Hexadecimal 268
Digital Circuits 274
Positive versus Negative Logic 274
Logic Gates 275
Black Boxes 276
Forms of Binary Data 277
Clocks 278
Counters 278
Digital Signals 281
Bits 281
Bytes 282
Baud 282
Examples of Data Speed 283
Analog-to-Digital Conversion 283
Digital-to-Analog Conversion 285
Serial versus Parallel 285
Digital Signal Processing 286
Digital Color 286
Quiz 292

CHAPTER 10 **Reality Remystified** **297**

The Illogic of Time 298
- The Light-Beam Conundrum 298
- Synchronize Your Watches! 299
- When Is a Second Not a Second? 300
- Traveling into the Future 303
- Traveling into the Past 304
- The Twin Paradox 304

The Illogic of Matter and Space 308
- Particles without End 308
- A Mad Professor's Monologue 309
- Order from Randomness 310
- The Ultimate State of Order 311
- Hyperspace 312
- Hypospace 313
- Time-Space 314
- Time as a Dimension 314
- Infinity-Space 316
- The Dwindling-Displacement Effect 317

The Illogic of Chaos 319
- Was Andrew "Due"? 319
- Slumps and Spurts 319
- Correlation, Coincidence, or Chaos? 320
- Scale-Recurrent Patterns 321
- The Maximum Unswimmable Time 322
- The Butterfly Effect 323
- Scale Parallels 324
- The Malthusian Model 325
- A Bumpy Ride 326
- What Is Randomness? 329

Final Exam *331*

Answers to Quizzes and Final Exam *363*

Suggested Additional Reading *365*

Index *367*

Preface

We wrote this book for people who want to learn about logic and related fields, or refresh their knowledge of those topics. The course can serve as a self-teaching guide or as a supplement in a classroom, tutored, or home-schooling environment.

With the exception of Chapter 10, each chapter ends with a multiple-choice quiz. You may (and should) refer back to the text when taking these quizzes. Because the quizzes are "open-book," some of the questions are difficult, but one choice is always best. When you're done with the quiz at the end of a chapter, give your list of answers to a friend. Have your friend tell you your score, but not which questions you got wrong. The correct choices appear in the back of the book. Stick with a chapter until you get all of the quiz answers correct.

The book concludes with a multiple-choice, "closed-book" final exam. Don't look back into the chapters when taking this test. A satisfactory score is at least 75 answers correct, but we suggest you shoot for 90. With the exam, as with the quizzes, have a friend tell you your score without letting you know which questions you missed. That way, you won't subconsciously memorize the answers. The questions have a format similar to those that you'll encounter in standardized tests.

We welcome suggestions for future editions!

Anthony Boutelle
Stan Gibilisco

Logic

DeMYSTiFieD®

chapter 1

Arguments, Validity, and Truth Tables

Logic involves the study of arguments and their use in reasoning. An *argument* tries to establish the truth of a certain statement, the *conclusion*, based on other statements, the *premises*. For example, if you say, "Nobody is perfect; therefore I am not perfect," the first part ("Nobody is perfect") constitutes the premise, and it gives a reason to accept the conclusion ("I am not perfect"). Few reasonable people would object to this mode of thought, but many arguments are far less clear-cut. What if we don't know that an argument actually *proves* a particular conclusion? What if one of the statements in an argument turns out false? Logic allows us to resolve most problems of this sort, and helps us to get an idea of how arguments "play out." Logic can also help us to know when people hope to fool us into accepting "conclusions" based on insufficient, irrelevant, or contradictory "evidence"!

CHAPTER OBJECTIVES

In this chapter, you will

- Learn what logic can do, and why we need it.
- Understand sentence forms and rules.
- Symbolize basic logical operations and sentences.
- Break sentences down into tables.
- Use truth tables to prove simple logical statements.
- Construct simple derivations.

What's Logic About?

When an argument functions as it should, and we know that the conclusion holds true on the strength of the premises, we have *sound* or *valid* reasoning (as opposed to *unsound* or *invalid* reasoning). Valid reasoning involves the way in which one sentence leads to another, regardless of whether the individual sentences hold true or not in the "real world." For example, we can construct a logically valid argument to the effect that if you are a sparrow, then you can fly (even though neither condition represents reality)! Logic provides us with mathematical certainty about a specific chain of reasoning. Pure logic, however, has nothing to do with the correctness or reality of any assumptions that we make.

Terminology

Logicians use many different terms to talk about arguments, but don't let them intimidate you! They're all talking about the same thing. We can call an argument an *inference* or a *deduction*, because we *infer* or *deduce* the conclusion from the premises. We might also say that the premises *entail* or *imply* the conclusion, or that the conclusion *follows from* the premises.

All the statements in an argument constitute *propositions*. In everyday spoken languages, we express statements as *sentences*. Simple *declarative sentences* lend themselves ideally to logic, for example, "All dogs are mammals." We have trouble assigning truth or falsity to questions, orders, or expressions of feeling, so such sentences do not lend themselves very well to use in logical arguments.

Logical Arguments

We can place verbal or "blogospheric" disputes—"arguments" in the everyday sense of the word—outside the bounds of formal logic. Such disputes rarely have much to do with deductive reasoning, and logic alone can rarely resolve them. We should also distinguish logical reasoning from the psychology that describes how people contemplate the nature of an argument. Logical validity constitutes an *objective*, not a *subjective*, measure of an argument's strength. Logical argument also differs from making a persuasive speech in support of a statement or idea, which often takes advantage of appeals to emotion and eloquent language (or even suspicious reasoning) in addition to (or instead of) valid inferences. Logic addresses only those statements about which we can speak or write with *absolute* certainty.

Even when dealing with facts, in many everyday circumstances we can do little more than guess at our conclusions, and we ought not to expect certainty from such diluted "reasoning." In a high-stakes criminal trial, the prosecutors must demonstrate the defendant's guilt "beyond a *reasonable* doubt." In logic, we demand total perfection; we must establish our conclusions "beyond *all* doubt." By expressing an inference as simply and clearly as possible and checking it for validity, we can make an "airtight" argument where true premises absolutely guarantee a true conclusion.

TIP ***Most arguments aren't as basic as deriving one conclusion from one premise. To deduce more complex, interesting things, we can string simple inferences together so that the conclusion of one step serves as a premise for the next step. As long as each link in the chain of inferences is valid, we can be sure that the whole argument is valid.***

What's Logic Good For?

Even though logic may seem abstract and divorced from practical concerns, it deals with truth at a general level, so it has broad applications. Mathematical proofs rely on logical inferences to establish significant final results known as *theorems*. Similar deductions apply in all fields of science. Computers rely on logic at a fundamental level; digital 1s and 0s work like "truths" and "falsehoods."

Traditionally, logic has constituted an important branch of philosophy, and logic remains central to philosophical analysis and argumentation. Logic can also apply to political science, debate, law, and rhetoric. Whenever we want to defend or critique an argument, we can use logic to classify the reasoning and test its validity.

Aside from all these uses, some logicians will tell you that logic has intrinsic value as a discipline in its own right, just as art has. Some people find beauty in the order and necessity of logical arguments. Others simply enjoy the "mental exercise" of working through a challenging logical proof.

Logical Form

Diverse arguments can share similar patterns of reasoning, so we don't have to start from scratch on every argument. This principle constitutes one of the most important insights of logical thinking!

An Example

Two arguments about vastly different subjects may share a single *logical form*. Consider the following:

- The ocean is made of water, and water is wet. Therefore, the ocean is wet.
- Candy is made of sugar, and sugar is sweet. Therefore, candy is sweet.

These two inferences share something in common, but it has nothing to do with the similarity between the ocean and candy. The statements *fit together* in the same way. We can characterize them both with the following argument form:

- Thing A is made of substance S, and substance S has property P. Therefore, thing A has property P.

We could fill in anything for A, S, and P without affecting the validity of the argument in the slightest.

When Truth Doesn't Matter

The form of an argument does not depend on the "real world" truth of the individual statements within it. The validity of the argument doesn't change even if we fill in the blanks with ridiculous statements. For example, consider the following argument:

- My house is made of figs, and figs can exist only in outer space. Therefore, my house exists in outer space.

The premises are both false, but they still entail the conclusion, which in this case is also false! The converse situation can also occur; sometimes we may come across an invalid argument whose conclusion happens to be true.

When Meaning Doesn't Matter

The truth of the statements isn't the only part of the argument that we can set aside to focus on the logical form; the meanings of the words may also lack relevance! That's why, for example, we can swap out the ocean, or candy, or figs in the foregoing arguments without having to reconsider the inference. In fact, we don't have to know what we're talking about at all! We can use nonsense words and remain confident that our reasoning holds sound. For example:

- Blurgs are made of gludds, and gludds are plishy. Therefore, blurgs are plishy.

Rules for Reasoning

We need a few guidelines in order to keep our arguments clear and consistent. Sometimes we will see such logical restrictions as self-evident "laws of thought." Even if you doubt this claim, it shouldn't stop you from following along. Think of these as the rules of the "logic game."

Identity

The *principle of identity* says that if a statement is true, then it is true (and if it's false, then it's false). Who can argue with that? In order for this rule to hold true all the time, however, we must use words consistently. The sentence "My father is tall" constitutes a true statement for some people and a false statement for others.

Still Struggling

The foregoing statement might seem confusing, but the confusion does not violate the principle of identity. The words "my father" can refer to a different man in each case, so the statements themselves differ! We must exercise caution whenever we work with words that depend on context (such as who speaks, or when they speak) and words with ambiguous meanings or multiple definitions.

Contradiction

The *principle of contradiction* tells us that no statement can hold true and false "at the same time." If an argument concludes both "It is the case that Peter is alive" and "It is not the case that Peter is alive," then something has gone wrong. If the argument uses words consistently as it should (so that "Peter" refers to the same man in both statements, and "alive" means the same thing), then the foregoing argument makes an incredible claim. Situations can arise in everyday discourse when we'll want to answer "yes and no," but in such cases, we'll inevitably find that at least one of the words has a double meaning.

True or False

The *law of excluded middle* maintains that all statements are either true or false. Traditional logic constitutes a *binary* (two-state) system. No middle values or half-truths exist. Even if we have no clue as to whether a statement holds true or false, we can safely assume that things must go one way or the other.

This principle may seem obvious to you, but it constitutes nothing more than an agreed-on convention for the simplest sorts of arguments. A few imaginative mathematicians and computer scientists have devised (and even applied) specialized forms of logic that allow for states other than true and false.

Sentence Forms

In the English language (and most other languages), we can break simple declarative sentences into two parts: *subject* and *predicate*. The subject comprises the main or only noun in the sentence, and it names what the sentence is about. The predicate provides us with information about the subject, such as a description of its qualities or expression of an action it takes.

Subject/Verb (SV) Statements

The simplest subject/predicate combination contains one noun and one verb. Consider the following sentences:

- Jack walks.
- Jill sneezes.
- The computer works.
- You shop.

These sentences all represent well-formed English propositions. They all have the *subject/verb* form in common, and they're all clear, unambiguous sentences. You can write SV (where S stands for "subject" and V stands for "verb") to symbolize each one of them. If their content seems spare (we aren't told where Jack is going, for example, or how he walks), that's because a subject/verb combination constitutes the most minimal structure that qualifies as a complete declarative sentence.

Subject/Verb/Object (SVO) Statements

Consider the following sentences:

- Jack walks his dog.
- Jill kicks the ball.
- You plant a tree.
- The ball hits the pavement.

Each of these sentences contains a noun (the subject) followed by a verb, and then a second noun that is acted upon by the subject and verb. We call this third part the *object*. All four of the above sentences have the form *subject/verb/object*. Let's abbreviate this form as SVO (where S stands for "subject," V stands for "verb," and O stands for "object"). Anything can serve as the subject or the object of a sentence: a person (Jack), an inanimate thing (ball), an abstract concept (fun), or whatever else we might imagine.

Subject/Linking Verb/Complement (SLVC) Statements

Now look at the following sentences:

- Jack is a person.
- Jill has a cold.
- You will be late.
- The dog was upset.

Once again, these sentences start with the main noun (subject). The final word of each sentence gives a description of the subject or another name for it; we call it the *complement*. The two words are joined by a verb, called the *linking verb*, usually a form of the verb "to be": am, is, are, was, were, or will be. (Some formal texts refer to a linking verb as a *copula*.) This sentence structure is called *subject/linking verb/complement*, or SLVC for short.

Symbols and Operations

In *propositional logic*, we will usually represent complete sentences by writing uppercase letters of the alphabet. You might say "It is raining outside," and represent this statement as the letter R. Someone else might add, "It's cold outside," and represent it as C. A third person might say, "The weather forecast calls for snow tomorrow," and represent it as S. Still another person might make the claim, "Tomorrow's forecast calls for sunny weather," and represent it as B (for "bright"; you've already used S).

Logical Negation (NOT)

When you write down a letter to stand for a sentence, you assert that the sentence holds true. So, for example, if Joanna writes down C in the above situation, she means to say "It is cold outside." You might disagree if you grew up in Alaska and Joanna grew up in Hawaii. You might say, "It's not cold outside." You would symbolize this statement by writing the letter C with a *negation* symbol in front of it.

In propositional logic, the drooping minus sign (¬) can represent negation, also called the *logical NOT operation*. Let's employ this symbol. Some texts use a tilde (~) to represent negation. Some use a minus sign (−). Some put a line over the letter representing the sentence; still others use an accent symbol. In our system, we would denote the sentence "It's not cold outside" by writing a drooping minus sign followed by the letter C; that is, by writing ¬C.

Now imagine that someone declares, "You are correct to say ¬C. In fact, I'd say that it's hot outside!" Let's use H to stand for "It's hot outside." If you give the matter a little bit of thought, you'll realize that H does not mean the same thing as ¬C. You've seen days that were neither cold nor hot. Our finicky atmosphere can produce in-between states such as "cool" (K), "mild" (M), and "warm" (W). There can exist no in-between condition, however, when it comes to the statements C and ¬C. In propositional logic, either it's cold or it's not cold. Either it's hot or it's not hot. Any given proposition is either true or it's false (not true). Of course, the "decision line" for temperature opinions will vary from person to person—but you should get the general idea!

Mathematicians have invented logical systems in which in-between states can and do exist. These schemes go by names such as *trinary logic* or *fuzzy logic*.

In trinary logic, three truth values exist: true, false, and "neutral." In fuzzy logic, we get a continuum of values, a smooth transition from "totally false" through "neutral" to "totally true." For now, let's stick to *binary logic*, in which any given proposition is either "purely false" or "purely true."

Logical Conjunction (AND)

Propositional logic doesn't bother with how words or phrases interact inside a sentence. However, propositional logic does involve the ways in which complete sentences interact with one another. We can combine sentences to make bigger ones, called *compound sentences*. The truth or falsity of a compound sentence depends on the truth or falsity of its component sentences, and on the ways in which we interconnect them.

Suppose someone says, "It's cold outside, and it's raining outside." Using the symbols above, we can write this as

C AND R

In *symbolic logic*, mathematicians use a symbol in place of the word AND. Several symbols appear in common usage, including the ampersand (&), the inverted wedge (∧), the asterisk (*), the period (.), the multiplication sign (×), and the raised dot (·). Let's use the inverted wedge. In that case, we write the above compound sentence as

$$C \wedge R$$

The formal term for the AND operation is *logical conjunction*. A compound sentence containing one or more conjunctions holds true if (but only if) both or all of its components are true. If any one of the components happens to be false, then the whole compound sentence is false.

Inclusive Logical Disjunction (OR)

Now imagine that a friend comes along and says, "You are correct in your observations about the weather. It's cold and raining. I have been listening to the radio, and I heard the weather forecast for tomorrow. It's supposed to be colder tomorrow than it is today. But it's going to stay wet. So it might snow tomorrow."

You say, "It will rain or it will snow tomorrow, depending on the temperature."

Your friend says, "It might be a mix of rain and snow together, if the temperature hovers near freezing."

"So we might get rain, we might get snow, and we might get both," you say.

"Of course. But the weather experts say we are certain to get precipitation of some sort," your friend says. "Water will fall from the sky tomorrow—maybe liquid, maybe solid, and maybe both."

Suppose that we let R represent the sentence "It will rain tomorrow," and we let S represent the sentence "It will snow tomorrow." Then we can say

S OR R

This statement constitutes an example of *inclusive logical disjunction*. Mathematicians use at least two different symbols to represent this operation: the plus sign (+) and the wedge (∨). Let's use the wedge. We can now write

$$S \vee R$$

A compound sentence in which both, or all, of the components are joined by inclusive disjunctions holds true if, but only if, at least one of the components is true. A compound sentence made up of inclusive disjunctions is false if, but only if, all the components are false.

Exclusive Logical Disjunction (XOR)

In the foregoing example, we refer to the *inclusive* OR *operation*. If all the components are true, then the whole sentence holds true. However, that's not the only way we can define disjunction. Once in awhile you'll encounter another logical operation called *exclusive* OR in which, if both or all of the components are true, the compound sentence is false.

If you say "It will rain or snow tomorrow" using the inclusive OR, then your statement will turn out true if the weather gives you a mix of rain and snow. But if you use the exclusive OR, you can't have a mix. The weather must produce either rain or snow, but not both, in order for the compound sentence to hold true.

TIP ***Some logicians call the exclusive OR operation the*** **either-or operation.** ***Engineers often abbreviate it as XOR; you'll encounter it if you do any in-depth study of digital electronic or computer systems. From now on, if you see the symbol for the OR operation (or the word "or" in a problem), you should assume that it means the inclusive OR operation,*** **not the exclusive OR operation.**

Logical Implication (IF/THEN)

Imagine that your conversation about the weather continues, getting more strange with each passing minute. You and your friend want to decide if you should prepare for a snowy day tomorrow, or conclude that you'll have to contend with nothing worse than rain and gloom.

"Does the weather forecast say anything about snow?" you ask.

"Not exactly," your friend says. "The radio newscaster said that there's going to be precipitation through tomorrow night, and that it's going to get colder tomorrow. I looked at my car thermometer as she said that, and the outdoor temperature was only a little bit above freezing."

"If we have precipitation, and if it gets colder, then it will snow," you say.

"Yes."

"Unless we get an ice storm."

"That won't happen."

"Okay," you say. "If we get precipitation tomorrow, and if it's colder tomorrow than it is today, then it will snow tomorrow." (This is a weird way to talk, but you're learning logic here, not the art of conversation. Logically rigorous conversation can sound bizarre, even in the "real world." Have you ever sat in a courtroom during a civil lawsuit between corporations?)

Let P represent the sentence, "There will be precipitation tomorrow." Let S represent the sentence "It will snow tomorrow," and let C represent the sentence "It will be colder tomorrow." In the above conversation, you made a compound proposition consisting of three sentences, as follows:

IF (P AND C), THEN S

Another way to write this is

(P AND C) IMPLIES S

In this context, "implies" means "always results in." In formal propositional logic, "X IMPLIES Y" means "If X, then Y." Symbolically, we write the above statement as

$$(P \wedge C) \rightarrow S$$

The arrow represents *logical implication,* also known as the *IF/THEN operation.* When we join two sentences with logical implication, we call the "implying" sentence (to the left of the arrow) the *antecedent*. In the above example, the antecedent is $(P \wedge C)$. We call the "implied" sentence (to the right of the arrow) the *consequent*. In the above example, the consequent is S.

Some texts use other symbols for logical implication, such as a "hook" or "lazy U opening to the left" (⊃) or a double-shafted arrow (⇒). Let's keep using the single-shafted arrow (→), making sure that it always points to the right.

Logical Equivalence (IFF)

Imagine that your friend declares, "If it snows tomorrow, then there will be precipitation and it will be colder." For a moment you hesitate, because that strikes you as an exceedingly strange way to make a statement about the weather. But you conclude that it makes perfect logical sense. Your friend has composed the following implication:

$$S \rightarrow (P \wedge C)$$

Now you and your friend agree that both of the following implications are valid:

$$(P \wedge C) \rightarrow S$$

and

$$S \rightarrow (P \wedge C)$$

You can combine these two statements into a logical conjunction, because you assert them both together "at the same time":

$$[(P \wedge C) \rightarrow S] \wedge [S \rightarrow (P \wedge C)]$$

When implications between two sentences hold valid "in both directions," we have an instance of *logical equivalence*. We can shorten the above statement to

$$(P \wedge C) \text{ IF AND ONLY IF } S$$

Mathematicians sometimes reduce the phrase "if and only if" to the single word "iff," so we can also write

$$(P \wedge C) \text{ IFF } S$$

Logicians use a double-headed arrow (↔) to symbolize logical equivalence. As with all the other logical operations, alternative symbols exist. Sometimes you'll see an equals sign (=), a three-barred equals sign (≡), or a double-shafted, double-headed arrow ($\Leftrightarrow$). Let's use the single-shafted, double-headed arrow ($\leftrightarrow$). We can now write the foregoing statement as

$$(P \wedge C) \leftrightarrow S$$

TIP ***When we want to establish logical equivalence, we must exercise care to ensure that the implication actually holds valid in both directions. Here's a situation in which logical implication holds in one direction but not in the other, but that limitation can easily escape us if we're careless. Consider this statement: "If it is overcast, then there are clouds in the sky." This statement always holds true. Suppose that we let O represent the sentence "It is overcast" and we let K represent the sentence "There are clouds in the sky." Then symbolically we have***

$$O \rightarrow K$$

If we reverse the sense of the implication, we obtain

$$K \rightarrow O$$

This sentence translates to, "If there are clouds in the sky, then it's overcast." This does not always hold true. We have all seen days or nights in which there were clouds in the sky, but there were clear spots too, so it wasn't overcast.

PROBLEM 1-1

Under what circumstances does the conjunction of several sentences have false truth value? Under what circumstances does the conjunction of several sentences hold true?

SOLUTION

The conjunction of several sentences is false if at least one of them is false. The conjunction of several sentences holds true if and only if every single one of them is true.

PROBLEM 1-2

Under what circumstances is the disjunction of several sentences false? Under what circumstances is the disjunction of several sentences true?

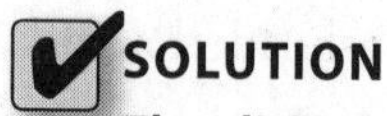

The disjunction of several sentences is false if and only if each and every one of them is false. The disjunction of several sentences holds true if at least one of them is true. In this context, the term *disjunction* refers to the inclusive OR operation.

Tables for Basic Operations

A *truth table* denotes all possible combinations of truth values for the variables in a compound sentence. The values for the individual variables appear in vertical columns.

Table for Logical Negation

The negation operation has the simplest truth table, because it operates on only one variable. Table 1-1 shows how logical negation works for a variable called X.

TABLE 1-1 Truth table for logical negation (NOT).

X	$\neg$X
F	T
T	F

Table for Conjunction

Let X and Y represent two logical variables. Conjunction $(X \wedge Y)$ produces results as shown in Table 1-2. This operation produces the truth value T when, but only when, both variables have value T. Otherwise, the operation produces the truth value F.

TABLE 1-2 Truth table for logical conjunction (AND).

X	Y	$X \wedge Y$
F	F	F
F	T	F
T	F	F
T	T	T

Tables for Logical Disjunction

Inclusive logical disjunction for two variables (X ∨ Y) breaks down as in Table 1-3A. We get T when either or both of the variables have truth value T. If both of the variables have truth value F, then we get F for the whole statement.

Table 1-3B shows the truth values for the exclusive logical disjunction operation (XOR). This operation gives us T if and only if both variables have opposite truth value (F-T or T-F); if the two variables agree (T-T or F-F), then the composite sentence has value F.

TABLE 1-3 Truth tables for both forms of logical disjunction. At A, inclusive (OR). At B, exclusive (XOR).

	A	
X	**Y**	**X ∨ Y**
F	F	F
F	T	T
T	F	T
T	T	T

	B	
X	**Y**	**X XOR Y**
F	F	F
F	T	T
T	F	T
T	T	F

Table for Logical Implication

A logical implication holds valid (has truth value T) whenever the antecedent is false. A logical implication is also valid if the antecedent and the consequent are both true. But implication has truth value F when the antecedent is true and the consequent is false. Table 1-4 shows the truth-value breakdown for logical implication.

TABLE 1-4 Truth table for logical implication (IF/THEN).

X	Y	$X \rightarrow Y$
F	F	T
F	T	T
T	F	F
T	T	T

Let's look at a "word-problem" example of logical implication that fails the "validity test." Let X represent the sentence, "The barometric pressure is falling fast." Let Y represent the sentence, "A tropical hurricane is coming." Consider the sentence

$$X \rightarrow Y$$

Now imagine that the barometric pressure is dropping faster than you've ever seen it fall. Therefore, variable X has truth value T. But suppose that you live in North Dakota, where tropical hurricanes never stray. Sentence Y has truth value F. Therefore, the implication "If the barometric pressure is falling fast, then a tropical hurricane is coming" has truth value F overall. In other words, it's invalid—in North Dakota, anyhow.

Table for Logical Equivalence

If X and Y represent logical variables, then X IFF Y has truth value T when both variables have value T, or when both variables have value F. If the truth values of X and Y differ, then X IFF Y has truth value F. Table 1-5 breaks the truth values down for logical equivalence.

TABLE 1-5 Truth table for logical equivalence (IFF).

X	Y	$X \leftrightarrow Y$
F	F	T
F	T	F
T	F	F
T	T	T

TIP *Note that the truth values for logical equivalence precisely oppose those for the exclusive OR operation (XOR). You can see this distinction when you compare the far right-hand columns of Tables 1-3B and 1-5.*

TIP *In logic, we can use an ordinary equals sign to indicate truth value. If we want to say that a particular sentence K holds true, for example, we can write K = T. If we want to say that a variable X always has false truth value, we can write X = F. But if we use the equals sign for this purpose, we must take care not to confuse its meaning with the meaning of the double-headed arrow that stands for logical equivalence. The equals sign tells us a characteristic of a particular sentence (truth or falsity); the double-headed arrow denotes an operation that we carry out between two sentences.*

A Quick Proof

Logicians define the truth values shown in Tables 1-1 through 1-4 by convention, using common sense. Arguably, we can use the same simple reasoning to get Table 1-5 for logical equivalence.

We might suppose that two logically equivalent statements must have identical truth values. How, we might ask, could common sense dictate anything else? We can back up our intuition, no matter how strong, by proving this fact based on the truth tables for conjunction and implication. Let's construct the proof in the form of a truth table.

Remember that $X \leftrightarrow Y$ means the same thing as $(X \rightarrow Y) \wedge (Y \rightarrow X)$. We can build up $X \leftrightarrow Y$ in steps, as shown in Table 1-6 as we proceed from left to right. The four possible combinations of truth values for sentences X and Y

TABLE 1-6 A truth-table proof that logically equivalent statements always have identical truth values.

X	Y	X → Y	Y → X	(X → Y) ∧ (Y → X)
F	F	T	T	T
F	T	T	F	F
T	F	F	T	F
T	T	T	T	T

appear in the first (left-most) and second columns. The truth values for X → Y appear in the third column, and the truth values for Y → X appear in the fourth column.

In order to get the truth values for the fifth (right-most) column, we can apply conjunction to the truth values in the third and fourth columns. The *complex logical operation* (also called a *compound logical operation* because it comprises combinations of the basic ones) in the fifth column has the same truth values as X ↔ Y in every possible case. Therefore, that compound operation does, in fact, constitute logical equivalence.

Q.E.D.

You've just seen a mathematical proof of the fact that for any two logical sentences X and Y, (X ↔ Y) = T when X and Y have the same truth value, and (X ↔ Y) = F when X and Y have different truth values. Sometimes, when mathematicians finish proofs, they write "Q.E.D." at the end. This sequence of letters constitutes an abbreviation of the Latin phrase *Quod erat demonstradum*. In English, it means "Which was to be demonstrated."

Precedence

When you read or construct logical statements, you should always do the operations within parentheses first. If you see multilayered combinations of sentences (called *nesting of operations*), then you should first use ordinary parentheses ( ), then brackets [], and then braces {}. Alternatively, you can use groups of plain parentheses inside each other, but if you do that, you had better ensure that you end up with the same number of left-hand parentheses and right-hand parentheses in the complete expression.

If you see an expression with no parentheses, brackets, or braces, you should go through the following steps in the order listed:

- Perform all the negations, going from left to right
- Perform all the conjunctions, going from left to right
- Perform all the disjunctions, going from left to right
- Perform all the implications, going from left to right
- Perform all the logical equivalences, going from left to right

We call this "operation hierarchy" *precedence of operations*, or simply *precedence*.

Examples of Precedence

Consider the following compound sentence, which might easily confuse anyone not familiar with the rules of precedence:

$$A \wedge \neg B \vee C \rightarrow D$$

Using parentheses, brackets, and braces to clarify this statement according to the rules of precedence, we can write

$$\{[A \wedge (\neg B)] \vee C\} \rightarrow D$$

Now consider the following compound sentence, which creates such a mess that we'll run out of grouping symbols if we use the "parentheses/brackets/braces" or "PBB" scheme:

$$A \wedge \neg B \vee C \rightarrow D \wedge E \leftrightarrow F \vee G$$

Using plain parentheses only, we can write it as

$$(((A \wedge (\neg B)) \vee C) \rightarrow (D \wedge E)) \leftrightarrow (F \vee G)$$

Still Struggling

Expressions such as the one shown above can confound even the most meticulous logician. When we count up the number of left-hand parentheses and the number of right-hand parentheses in the complete expression, we find six left-hand ones and six right-hand ones. Whenever we write, or read, a complicated logical sentence, we should always check to make sure that we have the proper balance of grouping symbols. It doesn't hurt to check complicated expressions two or three times!

PROBLEM 1-3

How many possible combinations of truth values exist for a set of four sentences, each of which can attain either the value T or the value F independently of the other three?

SOLUTION

For a group of four sentences, each of which can attain the value T or F independently of the other three, you can have 2^4 or 16 different combinations of truth values. If you think of F as the number 0 and T as the number 1, then you can find all the truth values of *n* independent sentences by counting up to 2^n in the binary numbering system. In this case, with $n = 4$, you would count as follows:

FFFF
FFFT
FFTF
FFTT
FTFF
FTFT
FTTF
FTTT
TFFF
TFFT
TFTF
TFTT
TTFF
TTFT
TTTF
TTTT

Proofs Using Truth Tables

If you claim that two compound sentences are logically equivalent, then you can prove that fact by showing that their truth tables produce identical results. Also, if you can show that two compound sentences have truth tables that produce identical results, then you can be sure that those two sentences are logically equivalent, as long as you account for all possible combinations of truth values. Following are several examples of simple proofs using truth tables.

TABLE 1-7 At A, statement of truth values for $X \wedge Y$. At B, statement of truth values for $Y \wedge X$. The outcomes are identical, demonstrating that they are logically equivalent.

A

X	Y	$X \wedge Y$
F	F	F
F	T	F
T	F	F
T	T	T

B

X	Y	$Y \wedge X$
F	F	F
F	T	F
T	F	F
T	T	T

Reversing the Order of a Conjunction

Tables 1-7A and 1-7B show that, for any two variables X and Y, the statement

$$X \wedge Y$$

is logically equivalent to the statement

$$Y \wedge X$$

We can write the foregoing theorem entirely in symbols as

$$X \wedge Y \leftrightarrow Y \wedge X$$

Grouping of Conjunctions

Tables 1-8A and 1-8B show that for any three variables X, Y, and Z, the statement

$$(X \wedge Y) \wedge Z$$

is logically equivalent to the statement

$$X \wedge (Y \wedge Z)$$

TABLE 1-8A Derivation of truth values for $(X \wedge Y) \wedge Z$. Note that the last two columns of this proof make use of a theorem that we've already proved.

X	Y	Z	$X \wedge Y$	$Z \wedge (X \wedge Y)$	$(X \wedge Y) \wedge Z$
F	F	F	F	F	F
F	F	T	F	F	F
F	T	F	F	F	F
F	T	T	F	F	F
T	F	F	F	F	F
T	F	T	F	F	F
T	T	F	T	F	F
T	T	T	T	T	T

In Table 1-8A, the logic between the last two columns makes use of the theorem that we proved in Table 1-7. When a theorem plays the role of a "subtheorem" in this way, we call it a *lemma*. We can write the foregoing final theorem entirely in symbols as

$$(X \wedge Y) \wedge Z \leftrightarrow X \wedge (Y \wedge Z)$$

TABLE 1-8B Derivation of truth values for $X \wedge (Y \wedge Z)$. The far-right-hand column has values that coincide with those in the far-right-hand column of Table 1-8A, demonstrating that the far-right-hand expressions in the top rows of both tables are logically equivalent.

X	Y	Z	$Y \wedge Z$	$X \wedge (Y \wedge Z)$
F	F	F	F	F
F	F	T	F	F
F	T	F	F	F
F	T	T	T	F
T	F	F	F	F
T	F	T	F	F
T	T	F	F	F
T	T	T	T	T

TABLE 1-9 At A, statement of truth values for X ∨ Y. At B, statement of truth values for Y ∨ X. The outcomes are identical, demonstrating that they are logically equivalent.

A		
X	**Y**	**X ∨ Y**
F	F	F
F	T	T
T	F	T
T	T	T

B		
X	**Y**	**Y ∨ X**
F	F	F
F	T	T
T	F	T
T	T	T

Reversing the Order of an Inclusive Disjunction

Tables 1-9A and 1-9B show that for any two variables X and Y, the statement

$$X \vee Y$$

is logically equivalent to the statement

$$Y \vee X$$

We can write the foregoing theorem entirely in symbols as

$$X \vee Y \leftrightarrow Y \vee X$$

Grouping of Inclusive Disjunctions

Tables 1-10A and 1-10B show that for any three variables X, Y, and Z, the statement

$$(X \vee Y) \vee Z$$

TABLE 1-10A Derivation of truth values for $(X \vee Y) \vee Z$. Note that the last two columns of this proof make use of a lemma that we've already proved.

X	Y	Z	$X \vee Y$	$Z \vee (X \vee Y)$	$(X \vee Y) \vee Z$
F	F	F	F	F	F
F	F	T	F	T	T
F	T	F	T	T	T
F	T	T	T	T	T
T	F	F	T	T	T
T	F	T	T	T	T
T	T	F	T	T	T
T	T	T	T	T	T

is logically equivalent to the statement

$$X \vee (Y \vee Z)$$

Our reasoning between the last two columns in Table 1-10A employs the theorem from Table 1-9 as a lemma. We can write the foregoing final theorem entirely in symbols as

$$(X \vee Y) \vee Z \leftrightarrow X \vee (Y \vee Z)$$

TABLE 1-10B Derivation of truth values for $X \vee (Y \vee Z)$. The far-right-hand column has values that coincide with those in the far-right-hand column of Table 1-10A, demonstrating that the far-right-hand expressions in the top rows of both tables are logically equivalent.

X	Y	Z	$Y \vee Z$	$X \vee (Y \vee Z)$
F	F	F	F	F
F	F	T	T	T
F	T	F	T	T
F	T	T	T	T
T	F	F	F	T
T	F	T	T	T
T	T	F	T	T
T	T	T	T	T

TABLE 1-11 At A, derivation of truth values for $X \rightarrow Y$. At B, derivation of truth values for $\neg Y \rightarrow \neg X$. The outcomes coincide, demonstrating that the two statements are logically equivalent.

A

X	Y	$X \rightarrow Y$
F	F	T
F	T	T
T	F	F
T	T	T

B

X	Y	$\neg Y$	$\neg X$	$\neg Y \rightarrow \neg X$
F	F	T	T	T
F	T	F	T	T
T	F	T	F	F
T	T	F	F	T

Reversing the Order of an Implication

Tables 1-11A and 1-11B show that for any two variables X and Y, the statement

$$X \rightarrow Y$$

is logically equivalent to the statement

$$\neg Y \rightarrow \neg X$$

We can write the foregoing theorem entirely in symbols as

$$X \rightarrow Y \leftrightarrow \neg Y \rightarrow \neg X$$

Ungrouping the Negation of a Conjunction

Tables 1-12A and 1-12B show that for any two variables X and Y, the statement

$$\neg(X \wedge Y)$$

TABLE 1-12 At A, derivation of truth values for $\neg(X \wedge Y)$. At B, derivation of truth values for $\neg X \vee \neg Y$. The outcomes coincide, demonstrating that the two statements are logically equivalent.

A			
X	Y	X ∧ Y	¬(X ∧ Y)
F	F	F	T
F	T	F	T
T	F	F	T
T	T	T	F

B				
X	Y	¬X	¬Y	¬X ∨ ¬Y
F	F	T	T	T
F	T	T	F	T
T	F	F	T	T
T	T	F	F	F

is logically equivalent to the statement

$$\neg X \vee \neg Y$$

We can write the foregoing theorem entirely in symbols as

$$\neg(X \wedge Y) \leftrightarrow \neg X \vee \neg Y$$

Ungrouping the Negation of an Inclusive Disjunction

Tables 1-13A and 1-13B show that for any two variables X and Y, the statement

$$\neg(X \vee Y)$$

is logically equivalent to the statement

$$\neg X \wedge \neg Y$$

We can write the foregoing theorem entirely in symbols as

$$\neg(X \vee Y) \leftrightarrow \neg X \wedge \neg Y$$

TABLE 1-13 At A, derivation of truth values for $\neg(X \vee Y)$. At B, derivation of truth values for $\neg X \wedge \neg Y$. The outcomes coincide, demonstrating that the two statements are logically equivalent.

A

X	Y	$X \vee Y$	$\neg(X \vee Y)$
F	F	F	T
F	T	T	F
T	F	T	F
T	T	T	F

B

X	Y	$\neg X$	$\neg Y$	$\neg X \wedge \neg Y$
F	F	T	T	T
F	T	T	F	F
T	F	F	T	F
T	T	F	F	F

Regrouping with Mixed Operations

Tables 1-14A and 1-14B show that for any three variables X, Y, and Z, the statement

$$X \wedge (Y \vee Z)$$

is logically equivalent to the statement

$$(X \wedge Y) \vee (X \wedge Z)$$

We can write the foregoing theorem entirely in symbols as

$$X \wedge (Y \vee Z) \leftrightarrow (X \wedge Y) \vee (X \wedge Z)$$

Truth Tables Have Power!

We can use truth tables to prove any statement in propositional logic, as long as it's valid, of course! Consider the rather arcane theorem

$$[(X \wedge Y) \rightarrow Z] \leftrightarrow [\neg Z \rightarrow (\neg X \vee \neg Y)]$$

TABLE 1-14A Derivation of truth values for $X \wedge (Y \vee Z)$.

X	Y	Z	$Y \vee Z$	$X \wedge (Y \vee Z)$
F	F	F	F	F
F	F	T	T	F
F	T	F	T	F
F	T	T	T	F
T	F	F	F	F
T	F	T	T	T
T	T	F	T	T
T	T	T	T	T

TABLE 1-14B Derivation of truth values for $(X \wedge Y) \vee (X \wedge Z)$. The far-right-hand column has values that coincide with those in the far-right-hand column of Table 1-14A, demonstrating that the far-right-hand expressions in the top rows of both tables are logically equivalent.

X	Y	Z	$X \wedge Y$	$X \wedge Z$	$(X \wedge Y) \vee (X \wedge Z)$
F	F	F	F	F	F
F	F	T	F	F	F
F	T	F	F	F	F
F	T	T	F	F	F
T	F	F	F	F	F
T	F	T	F	T	T
T	T	F	T	F	T
T	T	T	T	T	T

Tables 1-15A and 1-15B, taken together, prove that this theorem holds true for any three logical variables X, Y, and Z.

TABLE 1-15A Derivation of truth values for $(X \wedge Y) \rightarrow Z$.

X	Y	Z	$X \wedge Y$	$(X \wedge Y) \rightarrow Z$
F	F	F	F	T
F	F	T	F	T
F	T	F	F	T
F	T	T	F	T
T	F	F	F	T
T	F	T	F	T
T	T	F	T	F
T	T	T	T	T

TABLE 1-15B Derivation of truth values for $\neg Z \rightarrow (\neg X \vee \neg Y)$. The far-right-hand column has values that coincide with those in the far-right-hand column of Table 1-15A, demonstrating that the far-right-hand expressions in the top rows of both tables are logically equivalent.

X	Y	Z	$\neg X$	$\neg Y$	$\neg Z$	$\neg X \vee \neg Y$	$\neg Z \rightarrow (\neg X \vee \neg Y)$
F	F	F	T	T	T	T	T
F	F	T	T	T	F	T	T
F	T	F	T	F	T	T	T
F	T	T	T	F	F	T	T
T	F	F	F	T	T	T	T
T	F	T	F	T	F	T	T
T	T	F	F	F	T	F	F
T	T	T	F	F	F	F	T

PROBLEM 1-4

What, if anything, is wrong with the truth table shown in Table 1-16?

SOLUTION

Some of the entries in the far-right-hand column are incorrect.

TABLE 1-16 Truth table for Problems 1-4 and 1-5.

X	Y	Z	$X \vee Y$	$(X \vee Y) \wedge Z$
F	F	F	F	T
F	F	T	F	T
F	T	F	T	F
F	T	T	T	T
T	F	F	T	F
T	F	T	T	T
T	T	F	T	F
T	T	T	T	T

PROBLEM 1-5

What *single symbol* can we change to make Table 1-16 show a valid derivation?

SOLUTION

In the far-right-hand column header (top of the table), we can change the conjunction symbol ($\wedge$) to an implication symbol ($\rightarrow$) so that the header says $(X \vee Y) \rightarrow Z$.

QUIZ

You may refer to the text in this chapter while taking this quiz. A good score is at least 8 correct. Answers are in the back of the book.

1. **Consider the following argument: "The stones in Wyoming all fell from the moon. This stone came from Wyoming. Therefore, this stone fell from the moon." What purely logical flaws, if any, does this argument contain?**
   A. The first premise is ridiculous, so the argument must lack logical validity.
   B. We have no way of knowing whether the particular stone in question really came from Wyoming.
   C. We have no way of knowing how stones could fall from the moon.
   D. This argument contains no purely logical flaws.

2. **Imagine two propositions P and Q. Consider the compound statement A such that**

$$A \leftrightarrow P \text{ XOR } Q$$

   Now consider the compound statement B such that

$$B \leftrightarrow (P \leftrightarrow Q)$$

   Which of the following statements is logically valid?
   A. $A \rightarrow B$
   B. $A \leftrightarrow \neg B$
   C. $B \rightarrow A$
   D. $\neg A \leftrightarrow \neg B$

3. **According to the principle of identity,**
   A. no statement can be both true and false "at the same time."
   B. a direct contradiction can imply anything.
   C. if a statement is true, then it's true.
   D. every statement must identify a specific object or action.

4. **Consider the two statements, "Stan scribbles. Stan is a scribbler." Respectively, these sentences have the forms**
   A. SV and SLVC.
   B. SV and SVO.
   C. SLVC and SVO.
   D. SVO and SLVC.

5. **In a sound argument, we can have absolute confidence that**
   A. the conclusion leads to the premises beyond a reasonable doubt.
   B. the conclusion leads to the premises beyond any doubt whatsoever.
   C. the premises lead to the conclusion beyond a reasonable doubt.
   D. the premises lead to the conclusion beyond any doubt whatsoever.

6. **Based on the facts about logical conjunction between two variables that we've learned in this chapter, we can always do one of the following maneuvers and have confidence that the resulting statement is logically equivalent to the original statement. Which one?**
   A. We can switch the order of the variables.
   B. We can negate the first variable.
   C. We can negate the second variable.
   D. We can negate both variables.

7. **Based on the facts about inclusive logical disjunction between two variables that we've learned in this chapter, we can always do one of the following maneuvers and have confidence that the resulting statement is logically equivalent to the original statement. Which one?**
   A. We can switch the order of the variables.
   B. We can negate the first variable.
   C. We can negate the second variable.
   D. We can negate both variables.

8. **Consider the following compound sentence:**

$$\neg A \wedge B \wedge C \rightarrow D \wedge \neg E \leftrightarrow \neg F$$

   Which of the following character sequences represents a correct way to rewrite this statement with grouping symbols?
   A. $[\neg(A \wedge B \wedge C)] \rightarrow [D \wedge \neg(E \leftrightarrow \neg F)]$
   B. $\{[(\neg A) \wedge B] \wedge C\} \rightarrow \{[D \wedge (\neg E)] \leftrightarrow (\neg F)\}$
   C. $\{[(\neg A) \wedge B \wedge C] \rightarrow [D \wedge (\neg E)]\} \leftrightarrow (\neg F)$
   D. $\{[\neg(A \wedge B) \wedge C] \rightarrow [D \wedge (\neg E)]\} \leftrightarrow (\neg F)$

9. **A conjunction between the negations of two logical variables is always logically equivalent to**
   A. the negation of the exclusive disjunction between those two variables.
   B. the negation of the inclusive disjunction between those two variables.
   C. the negation of the implication from the second variable to the first.
   D. None of the above.

10. **An inclusive disjunction between the negations of two logical variables is always logically equivalent to**
   A. the negation of the exclusive disjunction between those two variables.
   B. the negation of the inclusive disjunction between those two variables.
   C. the negation of the implication from the second variable to the first.
   D. None of the above.

chapter 2

Propositional Logic

When we make an argument, we might "feel certain" about our reasoning, but in formal logic that's not good enough. We must produce a rigorous case that demonstrates why we *can't be wrong*, and we must formally *prove* our case to someone else. That's a tall order! If we really want to be sure about the validity of an argument, we need to write it down in a way that eliminates potential "gaps" between the steps, while at the same time maintaining clear language. We call this sure-fire way of expressing an argument a *formal system* or *formal logic*. The simplest way to learn proofs involves a formal system called *propositional logic*, also known as *sentential logic*, the *propositional calculus*, or the *sentential calculus*. *Propositional* and *sentential* refer to our subject matter, which comprises sentences. In logic, the term *calculus* doesn't refer to the mathematical discipline involving graphs and changing quantities; it reminds us that we can "calculate" proofs in a formal system of reasoning.

CHAPTER OBJECTIVES

In this chapter, you will

- String symbols together to construct propositional formulas.
- Learn the structural rules for sequents and arguments.
- Manipulate expressions containing logical operations.
- Apply fundamental laws of logic to prove simple theorems.
- Discover how to prove a theorem by deriving a contradiction.

A Formal System

To make a formal system in logic, we use symbols—a "logical language" without any of the subtle meanings and complicated grammar that make absolute certainty so hard to achieve in everyday speech. Because of the special logical language, some people call this system *symbolic logic*. Each symbol has only one meaning, and rules exist that tell us how we can use it. We can "translate" our arguments from English into the system when we want to prove something about them. We call the resulting string of symbols a *formula*.

Propositional Formulas (PFs)

We've been talking about sentences, but we haven't asked the basic question: What counts as a sentence in propositional logic? We can think of formulas made up of the symbols we have introduced that constitute mere gibberish, and that don't represent a definite proposition at all. Some examples follow:

$$)PQ(\neg$$

$$)P \wedge Q(\vee (R)$$

$$(P \wedge \leftrightarrow Q$$

Any string of "approved" symbols technically constitutes a formula, but it only expresses logical relationships when it follows the rules of syntax suggested as we introduced them. The formulas above don't express propositions, so we wouldn't know where to start assigning them truth values based on their variables. A formula that follows the rules (like most you'll encounter) is called a *propositional formula (PF)*, *propositional expression*, or *sentential formula*.

Let's recap our syntax rules to give a rigorous definition of what qualifies as a PF in the formal system we have set up:

- Any propositional variable (symbolized by a capital letter) is a PF.
- Any PF preceded by the symbol $\neg$ is a PF. For example, if X is a PF, then $\neg$X is a PF.
- Any PF joined by one of the symbols $\wedge$, $\vee$, $\rightarrow$, or $\leftrightarrow$ (known as *binary connectives* because they join together two propositions) to another PF, the whole thing inside parentheses, constitutes a PF. For example, if X is a PF, Y is a PF, and # is a binary connective, then (X # Y) is a PF.

- Any formula that does not satisfy at least one of the three conditions listed above does not constitute a PF.

All PFs constructed according to these rules include parentheses around everything but atomic propositions, including the whole formula as written. Too many parentheses can clutter a formula, so you can "drop" parentheses according to the established order of operations as an "informal" (but always unambiguous) shorthand. You can substitute the parentheses back in at any time to satisfy the strict definition of a PF above. Dropping unnecessary parentheses sometimes makes PFs appear simpler. You can use extra parentheses whenever you think that doing so will ensure that your reader correctly understands the formula.

Is This Mathematics?

We're using operators, variables, and formulas set off by parentheses in our formal system. If this notation reminds you of a course in mathematics, your hunch is well founded! Many of the conventions of formal logic were created with mathematics in mind. In fact, some people call symbolic logic mathematical logic. This doesn't mean propositional logic has anything to do with polygons or rational numbers, just that we want to achieve the same kind of rigor and certainty here that mathematicians use in, say, a proof in Euclidean geometry.

Contradiction

Some of our rules will involve the idea of a logical *contradiction*. This means what you might expect: a statement that disagrees with itself and therefore can't possibly hold true as a whole. We can't talk about contradiction in the everyday sense, though; we have to assign it a rigorous formal meaning that makes it as certain as the rest of the system.

Let C represent the sentence "It's cold outside," and let H represent the sentence "It's hot outside." You might at first think that the conjunction $C \wedge H$ constitutes a contradiction, but it really doesn't! We could disagree about the cutoff points for what we consider "hot" or "cold." We might define "in-between" states such as "warm"; this means that H does not negate C, strictly speaking. If we are to call a proposition a formal contradiction, that proposition must constitute the conjunction of a sentence with its own *exact* negation, as in $C \wedge \neg C$. Then we have no room for error; ¬C always has the opposite truth value from C, so we know for sure that one will be true and the other false (even though we might not know which is which). If we say "It's cold and it's not cold" or "It's hot and it's not hot," then we state a logical contradiction in the strictest formal sense.

Sequents

Any argument that we can prove in our system can be understood as a *sequent*, which means that we can summarize it as the set of propositions that act as premises and the single proposition that we prove from them. We will express all of our propositions as PFs, and we will divide the premises from the conclusions using a symbol called a *tee*, a *turnstile*, or an *assertion sign* ($\vdash$). We can read this out loud as "yields," "proves," or "therefore." We can read the expression

$$P \wedge Q \vdash Q$$

as "P and Q, therefore Q." We can also say that Q is *derivable* from P and Q. The assumption set on the left side is sometimes called the *antecedent*, and the conclusion on the right is called the *succedent*. (Be careful—the first part of a conditional proposition is also called an antecedent.) When we have multiple propositions in the antecedent, we'll separate them with commas, as we would do when writing a list; for example,

$$Q, R \vdash Q \wedge R$$

The order of the premises doesn't matter; we could have listed the premises in the last sequent as R, Q without producing a different sequent (and without changing the corresponding proof).

Still Struggling

We use the designation *sequent expression* because a written sequent does not constitute a PF or a proposition. Accordingly, $\vdash$ does not constitute a symbol of our formal language. As with truth assignments (and truth tables), we use sequents to talk about statements that constitute a part of the logical language, rather than to make statements within the language. The language that we use to discuss the propositional calculus, consisting of symbols and conventions used in this way, is called the logical *metalanguage*.

Backing Up the Argument

The purpose of having a logical system is to facilitate proofs of propositions without leaving anything to the imagination, so we always get valid results. Logicians have a formal way of writing down one proposition after another according to specific rules, and it gives anyone reading it all the pieces they need to follow the argument.

We write the sequent that we want to prove first; then we put down the propositions in our argument line by line beneath that initial sequent, expressing them all as PFs. In addition to the proposition itself, every line should include a line number for reference (on the left), a rule justifying the assertion, references showing how the rule fits in with the previous lines of the proof, and a tally of assumptions that we have employed to derive the proposition (all to the right of the PF). That gives us five columns in total. The reference column will indicate which propositions we're talking about (using the line numbers in the reference column), and the rule column shows what kind of deduction we're performing, using abbreviated names for the rules. This way, we can check each step of the proof for validity. We should always list the reference and assumption line numbers in numerical order.

The premise side of the sequent being proved is often introduced in the first line(s) of a proof, and the conclusion side should constitute the last line. In the assumptions column, we'll be able to see that the final proposition derives exclusively from the assumption(s) set out in the sequent.

Assumption (A)

A proof usually begins with an assumption and goes from there; the corresponding rule is the most basic one in the system. As you might imagine, we never deduce our assumptions from anything. Assumptions come "out of nowhere."

When we justify a proposition by *assumption* (abbreviated to an uppercase letter A), that assumption can comprise any claim we want to make, regardless of the other propositions that we've written down. We can introduce a proposition by assumption at any point in the proof. (We only need to make sure that the argument follows from the assumptions we've made—not that every assumption actually holds true in the "real world.") We don't have to reference any other lines from the proof as premises. We can simply write the letter A, and then put down the line number again under the assumptions column; the only assumption this line depends on is itself.

TABLE 2-1 The rule of assumption: P ⊢ P.

Line number	Proposition	Reference	Rule	Assumption
1	P		A	1

The simplest sequent for which we can write a "proof" is the fact that if we assume a preposition P, then we assert its truth. No one would dispute the fact that this makes a valid argument (even if it's the most trivial argument imaginable). Table 2-1 shows the formal demonstration. This constitutes a special case, because we can write the only premise and the conclusion with a single proposition, thereby proving the sequent in a single line.

Double Negation (DN)

The *rule of double negation* allows the negation operator to be introduced and eliminated based on the idea that two instances of ¬ in a row "cancel each other out." If someone says "It is not the case that it is not the case that it is cold outside," she's saying that it's cold outside; the statement ¬¬C is logically equivalent to the statement C.

When using the rule of double negation in a proof, you can abbreviate the rule as DN. You can use DN to add or take away pairs of negation symbols. Because DN transforms a single proposition into another logically equivalent one, you only need to reference that one line and copy over the same set of assumptions. Table 2-2 shows two simple examples.

TABLE 2-2 Two examples of double negation.

Introduction: P ⊢ ¬¬P

Line number	Proposition	Reference	Rule	Assumption
1	P		A	1
2	¬¬P	1	DN	1

Elimination: ¬¬P ⊢ P

Line number	Proposition	Reference	Rule	Assumption
1	¬¬P		A	1
2	P	1	DN	1

Conjunction Introduction (∧I)

If you want to derive a ∧-statement, you can use the *rule of conjunction introduction*, abbreviated as ∧I. This rule lets you construct a conjunction between two propositions that you've already derived. Write the line numbers for the two conjuncts in the reference column and combine their assumptions. All of the premises needed to establish the conjuncts are premises of the ∧-statement. Changing the order of a ∧-statement never changes its truth value, so you can put the resulting proposition in whatever order you wish. In the example shown by Table 2-3, you could derive Q ∧ P instead of P ∧ Q by following the same steps.

TABLE 2-3 The rule of conjunction introduction: $P, Q \vdash P \wedge Q$.

Line number	Proposition	Reference	Rule	Assumption
1	P		A	1
2	Q		A	2
3	P ∧ Q	1, 2	∧I	1, 2

Conjunction Elimination (∧E)

You will need a way to take apart ∧-statements as well. To do this, use the *rule of conjunction elimination*. This rule lets you derive the proposition on either side of the symbol ∧ by itself; the proposition will be the conjunct that you want to isolate. Cite the conjunction in the reference column, carry over all the same assumptions, and abbreviate the rule as ∧E. Just as with conjunction introduction, the order doesn't matter in the logical sense; you can "break off" either side of the conjunction. Table 2-4 illustrates an example.

TABLE 2-4 The rule of conjunction elimination: $P \wedge Q \vdash P$.

Line number	Proposition	Reference	Rule	Assumption
1	P ∧ Q		A	1
2	P	1	∧E	1

Disjunction Introduction (∨I)

You can use the *rule of disjunction introduction* to derive a ∨-statement on the basis of one or both of two propositions. That's the only line you must reference; you copy over the assumptions. The rule is abbreviated as ∨I. You can "add on" the other disjunct and ∨ symbol to either side of the premise. Though it may seem strange at first glance, you can introduce any PF you want as the unreferenced disjunct without adding any new assumptions, and you'll generate a valid deduction anyway. Table 2-5 shows an example.

TABLE 2-5 The rule of disjunction introduction: $P \vdash P \vee Q$.

Line number	Proposition	Reference	Rule	Assumption
1	P		A	1
2	P ∨ Q	1	∨I	1

Disjunction Elimination (∨E)

To get rid of a disjunction, you need the *rule of disjunction elimination*, abbreviated ∨E. This principle is a little trickier and more complex than the other inference rules that you've seen so far. It requires that you reference more lines. To establish that a proposition follows from a disjunction, you must prove that you can derive it from either disjunct by itself. When you want to prove something like this, you must introduce both disjuncts as assumptions, one at a time. You must therefore prove your desired conclusion twice: once following from each disjunct (along with any other assumptions that you make).

Don't worry about taking on "extra" premises for the proof as a whole. Such assumptions are temporary. You'll eliminate them before you finish the process. This is the first time that a rule has allowed you to "drop" an assumption, but this tactic constitutes a crucial part of doing proofs. If using assumptions provisionally and then discarding them were not allowed, you could never make any assumptions other than the premises given by the sequent.

The foregoing steps always take at least five lines to accomplish, even in the simplest case. On the line where you use the rule ∨E, you need to write five line numbers in the reference column, noting:

- The original ∨-statement
- The left-hand disjunct (as an assumption)

- The conclusion (derived from the left-hand disjunct)
- The right-hand disjunct (as an assumption)
- The conclusion (derived from the right-hand disjunct)

You should copy over all the assumptions that constitute premises for the disjunction at the start, as well as any extra assumptions (from elsewhere in the proof) that you might need to derive the conclusion along "either leg" of the argument. You use these assumptions to "test out" each disjunct as a part of the inference.

The sequence of logical steps shown in Table 2-6 uses ∨E to prove that the order of a disjunction doesn't matter (that is, Q ∨ P follows from P ∨ Q).

TABLE 2-6 Proof of P ∨ Q ⊢ Q ∨ P using disjunction elimination.

Line number	Proposition	Reference	Rule	Assumption
1	P ∨ Q		A	1
2	P		A	2
3	Q ∨ P	2	∨I	2
4	Q		A	4
5	Q ∨ P	4	∨I	4
6	Q ∨ P	1, 2, 3, 4, 5	∨E	1

Modus Ponens (MP)

An interesting rule known as *modus ponens* allows you to reason with conditional propositions. If you have derived an if/then statement and its antecedent, you can prove its consequent. The name of this rule derives from the Latin *modus ponendo ponens,* which translates to "the way that affirms by affirming." Let's abbreviate it as MP (some texts use MPP). Once in awhile, you'll hear a logician refer to the use of MP as *confirming the antecedent*.

You'll need two references to make an MP inference, corresponding to the conditional proposition and its antecedent. The concluding proposition on the MP line will be the consequent of the conditional proposition. To use MP, you combine the assumptions from both of the premises to get to the logical conclusion. Table 2-7 shows how the rule works in its most basic form.

TABLE 2-7 The rule of *modus ponens*: $P \rightarrow Q, P \vdash Q$.

Line number	Proposition	Reference	Rule	Assumption
1	$P \rightarrow Q$		A	1
2	P		A	2
3	Q	1, 2	MP	1, 2

Modus Tollens (MT)

The second way to argue from a conditional statement makes use of a rule called *modus tollens*. The term comes from the Latin expression *modus tollendo tollens*, which means "the way that denies by denying." Let's abbreviate it as MT (some texts use MTT). You may also see this argument form referred to as *denying the consequent*.

The rule of MT reasons that, if you negate the consequent of a conditional statement, you "automatically" negate the antecedent as well. To demonstrate MT, you must reference two lines (the conditional statement and the negation of its consequent) and combine their assumptions, as shown in Table 2-8.

TABLE 2-8 The rule of *modus tollens*: $P \rightarrow Q, \neg Q \vdash \neg P$.

Line number	Proposition	Reference	Rule	Assumption
1	$P \rightarrow Q$		A	1
2	$\neg Q$		A	2
3	$\neg P$	1, 2	MT	1, 2

Conditional Proof (CP) and Theorems

The rule of *conditional proof* (abbreviated CP) introduces the $\rightarrow$ symbol. This argument form allows you to "pull out" one of your assumed premises and turn it into an antecedent, making the proposition you derived from it into the consequent.

In order to use the rule of CP, assume the truth of the statement that you want to serve as the antecedent of the conditional, and then prove the proposition that you want to serve as the consequent, using the other methods of inference you've learned. On the line where you employ CP, reference the premise and conclusion that you will transform into a conditional proposition. The assumptions will be the same as the ones listed for the consequent, except that you'll drop the assumption corresponding to the antecedent. Unlike the other binary connectives, swapping the propositions from left to right (reversing the direction of the implication) makes a big difference in the meaning!

The rule of CP gives you enhanced power to get rid of superfluous assumptions. Sometimes, you can eliminate all of the assumptions from a proof! Logicians and mathematicians call a proposition that relies on no premises a *theorem*. The example in Table 2-9 represents the first time that we've seen this kind of sequent. Note that the antecedent side is blank. When reading off the symbol $\vdash$ with no premises on the left, don't read it as "yields," "proves," or "therefore." Instead, read it as "It is a theorem that..." In evaluative terms, theorems hold true under any assignment of truth values to atomic propositions. The proof in Table 2-9 translates to the statement "It is a theorem that $P \wedge Q \rightarrow P$."

TABLE 2-9 The rule of conditional proof: $\vdash P \wedge Q \rightarrow P$.

Line number	Proposition	Reference	Rule	Assumption
1	$P \wedge Q$		A	1
2	P		$\wedge$E	1
3	$P \wedge Q \rightarrow P$	1, 2	CP	

Reductio ad Absurdum (RA)

The rule of *reductio ad absurdum* allows you to derive a proposition by assuming its negation, and then showing that such an assumption leads to a *direct contradiction* (also called an *absurdity*)—the conjunction of some statement and its negation. The name translates from the Latin "reduction to the absurd"; we'll abbreviate it to RA (some texts use RAA).

Assuming the "opposite" of what you want to prove might at first seem like an illogical and counterintuitive thing to do, but it works! (Alternatively, you could think of it as assuming something in order to derive its negation.) Here's the idea: If you can derive a contradiction from a proposition, then that proposition cannot possibly hold true. If it isn't true, then its negation must hold true.

Start off by assuming the opposite of what you'd like to prove. If you want to prove P, then you must assume $\neg P$. Then, employ the other derivation techniques that you've learned to produce a contradiction of the form $Q \wedge \neg Q$, where Q can be any statement whatsoever. Once you have a contradiction, you can invoke RA; because you have a contradiction, you know that (at least) one of your assumptions needs to be denied in order to be consistent. You can "blame" the contradiction on any of the assumptions used to derive it, and then

TABLE 2-10 Proof of $P \rightarrow Q \wedge \neg Q \vdash \neg P$ using *reductio ad absurdum.*

Line number	Proposition	Reference	Rule	Assumption
1	$P \rightarrow Q \wedge \neg Q$		A	1
2	P		A	2
3	$Q \wedge \neg Q$	1, 2	MP	1, 2
4	$\neg P$	2, 3	RA	1

you can negate it based on the remaining assumptions. Finally, you drop the negated assumption, reducing the total assumption tally by one.

On the line where you negate the target proposition, reference both the line where you introduced it by assumption and the line on which a contradiction appeared (with the other line listed in the assumption column). You should copy over all of the assumptions cited on the contradiction line, other than the one that you intend to negate. Table 2-10 shows an example of RA at work.

Biconditional Introduction (↔I)

The rule of *biconditional introduction* says that if you can derive a conditional statement in "both directions" between two propositions, then you can put a ↔ symbol in between them. If you like, you can think of the biconditional as shorthand:

$$P \leftrightarrow Q \text{ means the same thing as } (P \rightarrow Q) \wedge (Q \rightarrow P)$$

To introduce a biconditional operator, reference the two conditional statements that show different directions of implication between the same two propositions, and then combine their assumptions. It doesn't matter which propositional formula (PF) you put on the left-hand side of the connective and which PF you put on the right-hand side. In proofs, you can symbolize this rule by writing ↔I. Table 2-11 illustrates ↔I in action.

TABLE 2-11 The rule of biconditional introduction: $P \rightarrow Q, Q \rightarrow P \vdash Q \leftrightarrow P$.

Line number	Proposition	Reference	Rule	Assumption
1	$P \rightarrow Q$		A	1
2	$Q \rightarrow P$		A	2
3	$Q \leftrightarrow P$	1, 2	↔I	1, 2

TABLE 2-12 The rule of biconditional elimination: $Q \leftrightarrow P \vdash P \rightarrow Q$.

Line number	Proposition	Reference	Rule	Assumption
1	$Q \leftrightarrow P$		A	1
2	$P \rightarrow Q$		↔E	1

Biconditional Elimination (↔E)

The rule of *biconditional elimination* allows you to "extract" a conditional proposition from a biconditional statement. To eliminate a biconditional operator, reference the line on which it appears and copy over the same assumptions. The result will be one of the two possible conditional statements, corresponding to one direction of implication in the original proposition. You can symbolize this rule in proofs by writing ↔E. Table 2-12 illustrates ↔E in action.

Substitution

Until now, we've treated propositional variables mainly in the expressions of simple declarative sentences. But variables can serve as stand-ins for any proposition, so we can substitute one for another whenever we want. For example, we've proved that

$$\neg\neg P \vdash P$$

so we can go back and construct another valid proof by substituting the letter Q every time P appears. Then we end up with the derivation

$$\neg\neg Q \vdash Q$$

We can substitute more complex propositions, too. For example, the statement

$$\neg\neg P \vdash P$$

allows us to derive, using substitution, the more complex statement

$$\neg\neg(Q \rightarrow R) \vdash (Q \rightarrow R)$$

Logical arguments depend only on the form and relationships of propositions, so all arguments of this form remain valid regardless of what statements we "plug in."

TIP *From the principle of substitution, we know that we can rely on the inference rules to apply to any PF, even though we've only shown them for P, Q, and R. Proofs in pure logic hold true in general. But we had better watch out! If we want substitution to work, we must treat every instance in exactly the same way. If we accidentally change the form (by, say, substituting Q for some instances of P and R for other instances of P), then we lose the validity of the PF, sequent or proof.*

PROBLEM 2-1

Use the inference rules to prove the sequent $P \rightarrow (Q \rightarrow R) \vdash Q \rightarrow (P \rightarrow R)$.

SOLUTION

See Table 2-13.

TABLE 2-13 Proof of the sequent $P \rightarrow (Q \rightarrow R) \vdash Q \rightarrow (P \rightarrow R)$.

Line number	Proposition	Reference	Rule	Assumption
1	$P \rightarrow (Q \rightarrow R)$		A	1
2	P		A	2
3	$Q \rightarrow R$	1, 2	MP	1, 2
4	Q		A	4
5	R	3, 4	MP	1, 2, 4
6	$P \rightarrow R$	2, 5	CP	1, 4
7	$Q \rightarrow (P \rightarrow R)$	4, 6	CP	1

PROBLEM 2-2

Prove the sequent $\neg P \rightarrow P \vdash P$.

SOLUTION

See Table 2-14.

TABLE 2-14 Proof of the sequent $\neg P \rightarrow P \vdash P$.

Line number	Proposition	Reference	Rule	Assumption
1	$\neg P \rightarrow P$		A	1
2	$\neg P$		A	2
3	P	1, 2	MP	1, 2
4	$P \wedge \neg P$	2, 3	∧I	1, 2
5	$\neg\neg P$	2, 4	RA	1
6	P	5	DN	1

PROBLEM 2-3

Prove the theorem $\vdash P \vee Q \rightarrow (P \vee R) \vee (Q \vee R)$.

SOLUTION

See Table 2-15.

TABLE 2-15 Proof of the theorem $\vdash P \vee Q \rightarrow (P \vee R) \vee (Q \vee R)$.

Line number	Proposition	Reference	Rule	Assumption
1	$P \vee Q$		A	1
2	P		A	2
3	$P \vee R$	2	∨I	2
4	$(P \vee R) \vee (Q \vee R)$	3	∨I	2
5	Q		A	5
6	$Q \vee R$	5	∨I	5
7	$(P \vee R) \vee (Q \vee R)$	6	∨I	5
8	$(P \vee R) \vee (Q \vee R)$	1, 2, 4, 5, 7	∨E	1
9	$P \vee Q \rightarrow (P \vee R) \vee (Q \vee R)$	1, 8	CP	

PROBLEM 2-4

Prove the sequent $\neg P \rightarrow Q, \neg Q \vdash P \vee R$.

SOLUTION

See Table 2-16.

TABLE 2-16 Proof of the sequent $\neg P \rightarrow Q, \neg Q \vdash P \vee R$.

Line number	Proposition	Reference	Rule	Assumption
1	$\neg P \rightarrow Q$		A	1
2	$\neg Q$		A	2
3	$\neg\neg P$	1, 2	MT	1, 2
4	P	3	DN	1, 2
5	$P \vee R$	4	∨I	1, 2

Laws

Logical operators obey certain rules known as *laws*. Several of the most basic and helpful laws of propositional logic follow. You can verify the validity of any law using a truth table. Do you recognize some of these tables from examples in the last chapter?

Law of Contradiction

As we've seen already, a formal contradiction always turns out false. We call this the *law of contradiction*, symbolized as follows:

$$(P \wedge \neg P) = F$$

Table 2-17 illustrates the law of contradiction.

TABLE 2-17 Truth table for the law of contradiction.

P	$\neg P$	$P \wedge \neg P$
T	F	F
F	T	F

Law of Excluded Middle

The *law of excluded middle* states that every proposition is either true or false. We can express this law symbolically as

$$(P = T) \vee (\neg P = T)$$

TABLE 2-18 Truth table for the law of excluded middle.

P	¬P	P ∨ ¬P
T	F	T
F	T	T

Logicians usually write it more simply as follows:

$$(P \lor \neg P) = T$$

Table 2-18 illustrates the law of excluded middle.

Law of Double Negation

The inference rule for double negation (DN) depends on the logical equivalence of a proposition and its double-negation.

$$\neg\neg P \leftrightarrow P$$

Table 2-19 illustrates the law of double negation.

TABLE 2-19 Truth table for the law of double negation.

P	¬P	¬ ¬P
T	F	T
F	T	F

Commutative Laws

Order doesn't matter for some operators, meaning that you can swap the propositions on both sides, and the result remains logically equivalent to the original. These properties give us the so-called *commutative laws*. In propositional logic, three commutative laws exist, as follows:

- Commutative law of conjunction: $(P \land Q) \leftrightarrow (Q \land P)$
- Commutative law of disjunction: $(P \lor Q) \leftrightarrow (Q \lor P)$
- Commutative law of the biconditional: $(P \leftrightarrow Q) \leftrightarrow (Q \leftrightarrow P)$

TABLE 2-20 Truth tables for the commutative laws.

A			
Commutative law of conjunction			
P	**Q**	**$P \wedge Q$**	**$Q \wedge P$**
T	T	T	T
T	F	F	F
F	T	F	F
F	F	F	F

B			
Commutative law of disjunction			
P	**Q**	**$P \vee Q$**	**$Q \vee P$**
T	T	T	T
T	F	T	T
F	T	T	T
F	F	F	F

C			
Commutative law of the biconditional			
P	**Q**	**$P \leftrightarrow Q$**	**$Q \leftrightarrow P$**
T	T	T	T
T	F	F	F
F	T	F	F
F	F	T	T

Tables 2-20A, 2-20B, and 2-20C verify the commutative laws of conjunction, disjunction, and the biconditional, respectively. Some logicians refer to the commutative laws as the *commutative properties*.

Associative Laws

When two conjunctions join three propositions, it doesn't matter how you group the variables. Suppose that you have the following informal PF:

$$P \wedge Q \wedge R$$

When writing in the parentheses, you can group it in either of two logically equivalent ways:

$$(P \wedge Q) \wedge R$$

or

$$P \wedge (Q \wedge R)$$

The same holds for disjunctions and biconditionals, so we have the three following *associative laws*.

- Associative law of conjunction: $[(P \wedge Q) \wedge R] \leftrightarrow [P \wedge (Q \wedge R)]$
- Associative law of disjunction: $[(P \vee Q) \vee R] \leftrightarrow [P \vee (Q \vee R)]$
- Associative law of the biconditional: $[(P \leftrightarrow Q) \leftrightarrow R] \leftrightarrow [P \leftrightarrow (Q \leftrightarrow R)]$

TIP ***Exercise caution with the associative laws. They only work when both operators in a three-statement PF are of the same type. For example, the following statement*** **does not** ***hold true in general:***

$$(P \vee Q) \wedge R \leftrightarrow P \vee (Q \vee R)$$

Check it with a truth table if you like. Tables 2-21A, 2-21B, and 2-21C verify the associative laws of conjunction, disjunction, and the biconditional, respectively. Some logicians call these laws the **associative properties.**

Law of Implication Reversal

The principle of *modus tollens*, which we learned earlier in this chapter, depends on the *law of implication reversal*, also known as the *law of the contrapositive*:

$$P \rightarrow Q \text{ is equivalent to } \neg Q \rightarrow \neg P$$

Table 2-22 demonstrates the law of implication reversal.

TABLE 2-21 Truth tables for the associative laws.

A

Associative law of conjunction

P	Q	R	$P \wedge Q$	$Q \wedge R$	$(P \wedge Q) \wedge R$	$P \wedge (Q \wedge R)$
T	T	T	T	T	T	T
T	T	F	T	F	F	F
T	F	T	F	F	F	F
T	F	F	F	F	F	F
F	T	T	F	T	F	F
F	T	F	F	F	F	F
F	F	T	F	F	F	F
F	F	F	F	F	F	F

B

Associative law of disjunction

P	Q	R	$P \vee Q$	$Q \vee R$	$(P \vee Q) \vee R$	$P \vee (Q \vee R)$
T	T	T	T	T	T	T
T	T	F	T	T	T	T
T	F	T	T	T	T	T
T	F	F	T	F	T	T
F	T	T	T	T	T	T
F	T	F	T	T	T	T
F	F	T	F	T	T	T
F	F	F	F	F	F	F

C

Associative law of the biconditional

P	Q	R	$P \leftrightarrow Q$	$Q \leftrightarrow R$	$(P \leftrightarrow Q) \leftrightarrow R$	$P \leftrightarrow (Q \leftrightarrow R)$
T	T	T	T	T	T	T
T	T	F	T	F	F	F
T	F	T	F	F	F	F
T	F	F	F	T	T	T
F	T	T	F	T	F	F
F	T	F	F	F	T	T
F	F	T	T	F	T	T
F	F	F	T	T	F	F

TABLE 2-22 Truth table for the law of implication reversal.

P	Q	¬P	¬Q	P → Q	¬Q → ¬P
T	T	F	F	T	T
T	F	F	T	F	F
F	T	T	F	T	T
F	F	T	T	T	T

DeMorgan's Laws

Two rules of logic show an interesting relationship among conjunction, disjunction, and negation. They're called *DeMorgan's laws,* and we can state them as follows:

- DeMorgan's law for conjunction: $\neg(P \wedge Q) \leftrightarrow \neg P \vee \neg Q$
- DeMorgan's law for disjunction: $\neg(P \vee Q) \leftrightarrow \neg P \wedge \neg Q$

Tables 2-23A and 2-23B verify these principles.

TABLE 2-23 Truth tables for DeMorgan's laws.

A

DeMorgan's law for conjunction

P	Q	¬P	¬Q	P ∧ Q	¬(P ∧ Q)	¬P ∨ ¬Q
T	T	F	F	T	F	F
T	F	F	T	F	T	T
F	T	T	F	F	T	T
F	F	T	T	F	T	T

B

DeMorgan's law for disjunction

P	Q	¬P	¬Q	P ∨ Q	¬(P ∨ Q)	¬P ∧ ¬Q
T	T	F	F	T	F	F
T	F	F	T	T	F	F
F	T	T	F	T	F	F
F	F	T	T	F	T	T

Distributive Laws

Another relationship involving conjunction and disjunction is described by the *distributive laws*. The first, the *distributive law of conjunction over disjunction*, states that a conjunction operator can be "distributed" over two disjuncts as follows:

$$P \wedge (Q \vee R) \leftrightarrow (P \wedge Q) \vee (P \wedge R)$$

The *distributive law of disjunction over conjunction* switches the roles of the two operators:

$$P \vee (Q \wedge R) \leftrightarrow (P \vee Q) \wedge (P \vee R)$$

Tables 2-24A and 2-24B demonstrate these principles, which some authors refer to as the *distributive properties*.

Interderivability ($\dashv\vdash$), Deriving Laws, and Theorem Introduction (TI)

When you have proved "both directions" of a sequent between two propositions (or sets of propositions), they are called *interderivable*, which implies logical equivalence. The symbol for interderivability comprises a double assertion sign, which we write as two "tee" symbols reversed and placed back to back ($\dashv\vdash$). Establishing interderivability requires two proofs, one for deriving the sequent "each way."

The laws we've shown using truth tables can also be proved using the derivation methods described in this chapter. For example, we executed the proof for the rule of disjunction elimination for the sequent

$$P \vee Q \vdash Q \vee P$$

A simple substitution of propositional variables allows us to see that

$$Q \vee P \vdash P \vee Q$$

From the two sequents above, we can conclude the sequential expression of the commutative principle

$$P \vee Q \dashv\vdash Q \vee P$$

We can also prove a version of the laws as theorems within the propositional calculus. For the commutative law of disjunction, a step using the rule of

TABLE 2-24 Truth tables for the distributive laws.

A							
Distributive law of conjunction over disjunction							
P	**Q**	**R**	**Q ∨ R**	**P ∧ Q**	**P ∧ R**	**P ∧ (Q ∨ R)**	**(P ∧ Q) ∨ (P ∧ R)**
T	T	T	T	T	T	T	T
T	T	F	T	T	F	T	T
T	F	T	T	F	T	T	T
T	F	F	F	F	F	F	F
F	T	T	T	F	F	F	F
F	T	F	T	F	F	F	F
F	F	T	T	F	F	F	F
F	F	F	F	F	F	F	F

B							
Distributive law of disjunction over conjunction							
P	**Q**	**R**	**Q ∧ R**	**P ∨ Q**	**P ∨ R**	**P ∨ (Q ∧ R)**	**(P ∨ Q) ∧ (P ∨ R)**
T	T	T	T	T	T	T	T
T	T	F	F	T	T	T	T
T	F	T	F	T	T	T	T
T	F	F	F	T	T	T	T
F	T	T	T	T	T	T	T
F	T	F	F	T	F	F	F
F	F	T	F	F	T	F	F
F	F	F	F	F	F	F	F

conditional proof can turn the sequents above into implication statements with no assumptions, and then we can combine them into the biconditional theorem

$$\vdash P \vee Q \leftrightarrow Q \vee P$$

When we state the above-described laws as theorems of the propositional language, we can use them in proofs by substituting in the relevant PF. For example, at any point in a proof, we can introduce the law of excluded middle and cite the rule of *theorem introduction* (TI), an example of which we see summarized in Table 2-25. By definition, theorems rely on no premises, so we need no references or assumptions.

TABLE 2-25 An example of theorem introduction on a line in a proof.

Line number	Proposition	Reference	Rule	Assumption
#	$(Q \vee P) \vee \neg(Q \vee P)$		TI	

You can employ any of the laws you've seen so far, as well as any theorem you might have previously proved in the course of your logical wanderings, in a theorem introduction.

QUIZ

You may refer to the text in this chapter while taking this quiz. A good score is at least 8 correct. Answers are in the back of the book.

1. **Which of the following portrays a correct way to read the symbol ⊢?**
   I. "...therefore..."
   II. "It is a theorem that...
   III. "...if and only if..."
   IV. "...yields..."
   V. "...is a tautology."

   A. I, II, and IV are correct.
   B. Only II is correct.
   C. I and V are correct.
   D. All of the above are correct.

2. **Suppose you observe that "It is not sunny outside, and it is not hot outside." Your friend says, "It is not the case that it is sunny or hot today." These two sentences are logically equivalent, providing an example of**
   A. one of DeMorgan's laws.
   B. the law of double negation.
   C. the law of implication reversal.
   D. no law; the two statements are not equivalent.

3. **Which of the following statements accurately describes the proof shown in Table 2-26?**
   A. The table constitutes a valid proof of the sequent.
   B. The proof contains errors in the assumption column, so it's invalid.
   C. The proof constitutes a valid deduction, but it does not prove the sequent.
   D. The proof contains errors in the rule column, so it's invalid.

TABLE 2-26 Proof of the sequent P ∨ Q, R ⊢ (P ∧ R) ∨ (Q ∧ R).

Line number	Proposition	Reference	Rule	Assumption
1	P ∨ Q		A	1
2	R		A	2
3	P		A	3
4	P ∧ R	2, 3	∧I	2, 3
5	(P ∧ R) ∨ (Q ∧ R)	4	∨I	2, 3
6	Q		A	6
7	Q ∧ R	2, 6	∧I	2, 6
8	(P ∧ R) ∨ (Q ∧ R)	7	∨I	2, 6

4. What, if anything, can be done to correct the proof in Question 3?

A. The antecedent of the sequent should be changed to $P \vee Q, P \wedge R$.
B. The rule at step 7 should be changed to disjunction elimination ($\vee$E).
C. An extra step using disjunction elimination ($\vee$E) should be added at the end.
D. Nothing needs to be done. It is correct as it is.

5. Suppose someone says to you, "If the sun is shining, it is hot outside. But it is not hot outside, so the sun must not be shining." This is an example of

A. one of DeMorgan's laws.
B. the distributive law of conjunction over disjunction.
C. the law of implication reversal.
D. no law; your friend's reasoning is invalid.

6. Suppose that a mathematician says, "A number is *rational* if and only if it equals the *ratio* of two integers." You respond by saying, "If I take an integer and divide it by another integer, I'll end up with a rational number." You can get this conclusion straightaway from the mathematician's claim by invoking

A. the rule of biconditional elimination.
B. *modus tollens*.
C. *modus ponens*.
D. the law of the contrapositive.

7. If we formulate a PF, no matter how complicated, we know that it must be either true or false based on the law of

A. implication reversal.
B. contradiction.
C. excluded middle.
D. interderivability.

8. Which of the following constitute PFs according to the rigorous definition of PF construction?

I. $P \rightarrow (Q \rightarrow R))$
II. $(P \wedge Q) = T$
III. $(P \wedge (\neg P))$
IV. $)PQ(\neg$
V. $\vdash ((P \wedge Q) \rightarrow P)$

A. I, III, and V are PFs.
B. Only III is a PF.
C. All but IV are PFs.
D. None of the above is a PF.

9. When using the rule of *modus ponens* (MP), we must copy over

A. the assumptions from the conditional statement and the antecedent.
B. the assumptions from the conditional statement only.
C. the assumptions from the conditional statement and the consequent.
D. nothing; modus ponens requires no assumptions.

10. If you say "It is hot and it is sunny, or it is hot and it is cloudy," the distributive law of conjunction over disjunction allows you to conclude

A. nothing, because the law does not properly apply to this statement.
B. the equivalent statement "It is not cloudy and it is sunny, or it is cloudy and not sunny."
C. the equivalent statement "If it is cloudy or sunny, then it is hot."
D. the equivalent statement "It is hot outside, and sunny or cloudy."

chapter 3

Predicate Logic

Propositional proofs and truth tables allow us to prove things about relationships between whole sentences, but for some arguments we must break sentences down and see their internal structure—how the constituent parts of a proposition work together. To do this kind of proof rigorously, we can use *predicate logic*, also called the *predicate calculus*. Like the propositional calculus, it's a formal system.

CHAPTER OBJECTIVES

In this chapter, you will

- Render complex sentences in symbolic form.
- Compare universal and existential quantifiers.
- Learn how quantifiers relate.
- Construct well-formed formulas.
- Work with two-part relations and syllogisms.
- Execute proofs in predicate logic.

Symbolizing Sentence Structure

In order to unambiguously construct statements and execute proofs that depend on internal sentence structure, we must introduce symbols that allow us to rigorously talk or write about them.

Formal Predicates

In formal logic, we represent *predicates* as capital letters of the alphabet called *predicate letters*. For example, we might write W for "walks" and J for "jumps." As with the variables of propositional logic, we can choose any letter to stand for a predicate, but we must avoid using the same letter for two different predicates in a single proof, derivation, or argument. (Otherwise we'll introduce ambiguity.)

Formal Subjects: Names

We represent the *subjects* of sentences as italicized lowercase letters called *subject letters*. A noun or proper name constitutes a simple, and common, subject. For instance, we might translate the proper name "Jack" to the letter *j* and the noun "science" to the letter *s*. When possible, we'll use the initial letter of the noun we want to symbolize, while making sure not to assign a particular letter more than once in a single proof, derivation, or argument.

Quite often, we'll want to construct sentences with the noun "any given person" or "a certain triangle" instead of naming a particular person or thing. We call this type of noun an *arbitrary name* (or an *arbitrary constant*) and symbolize it with an italicized, lowercase letter, just as we do with specific subjects. By convention, we'll start from the beginning of the alphabet (*a*, *b*, *c*, ...) when assigning these symbols. We can refer to proper and arbitrary nouns generally as *names*, *terms*, or *constants*.

Formal Subjects: Variables

When we need a "placeholder" to stand for a constant in a proposition, we will use a *logical variable*, also known as an *individual variable*. (We'll rarely run any risk of confusing it with a propositional variable, so we can simply call it a *variable*.) We symbolize variables, like names, with lowercase italicized letters.

The traditional letter choices are *x*, *y*, and *z*—just like the letters that represent common variables in mathematics. If we need more letters for variables, we can also use *u*, *v*, and *w*.

Predicate Sentence Formulas

To form a complete predicate statement, we can place a predicate letter next to a name or an individual variable, writing the predicate first. For example, if someone says "Roger eats beef," we can write it as E*r* (with E symbolizing "eats beef" and *r* symbolizing "Roger"). We call this form an *elementary sentence* (also known as an *atomic sentence* or *singular proposition*), because it constitutes the simplest form for a complete statement in the predicate calculus.

TIP ***Even though they possess internal structures (unlike propositional variables), atomic sentences are the simplest elements in the formal system of predicate logic to which we can assign truth values.***

PROBLEM 3-1

Translate the following sentences into formal predicate propositions using predicate letters and names:

- **Jill has a cold.**
- **You will arrive late.**
- **Ruth lives in America.**
- **I walked.**

SOLUTION

We replicate the sentences below, along with letters assigned as their symbols and translated into predicate formulas. You might want to assign different letters to the names and predicates, but the complete statements should have the same form as the ones below, no matter what particular letters you choose:

- **Jill (*l*) has a cold (C).** **C*l***
- **You (*u*) will arrive late (L).** **L*u***
- **Ruth (*r*) lives in America (A).** **A*r***
- **I (*i*) walked (W).** **W*i***

More Predicate Sentences

Not every predicate formula contains a single predicate and a single constant (or stand-in variable). If we want to symbolize "Jack walks to work" and "Ruth walks home," we could simply write K*j* and N*r*, but we can't see from this two-part structure that both Jack and Ruth are walking (as opposed to, say, running, riding bicycles, or driving cars). In order to symbolize this level of structural detail, we need a type of predicate called a *relation*. Relations apply to more than one constant (or variable), and they allow us to express a relationship between two things (instead of a property of only one thing). For this reason, logicians sometimes refer to relations as *dyadic* (two-part) and *polyadic* (many-part) *predicates*.

We can analyze sentences of this more complicated sort as relations with constants symbolizing both the subject of the sentence (Jack or Ruth) and the destination (school or home, respectively). To write a relation as a formula, we use a predicate letter followed by two names. For example, we might symbolize "Jack walks to school" and "Ruth walks to her house" as W*js* and W*rh*, respectively, where W stands for "__ walks to __." We can see from these formulas that the predicate has the same form in both cases, suggesting the analogy "*j* is to *s* as *r* is to *h*" (even if we don't know what the symbols stand for). Note that when we put more than one constant into a formula with a predicate letter, we must make sure that we put the symbols in the correct order. We don't want to end up with a formula that says "School walks to Jack!"

Most predicates that you'll likely see constitute one-place properties or two-place relations. But in theory, a predicate can take any finite number of constants. For example, we could make a three-place relation (using the letter assignments we established before for the names), symbolizing "Jack and Ruth are Jill's parents" by writing P*jrl*.

TIP ***Occasionally, we'll encounter a predicate that doesn't apply to any names at all. In a situation like this, we have an atomic sentence consisting of one predicate letter by itself; the resulting proposition constitutes an atomic propositional variable, and we can symbolize it the same way as we do in the propositional calculus (as a capital letter). We can consider a predicate accompanied by any finite number of names to constitute an atomic sentence, as long as we have the correct number of names for that specific predicate.***

Identity (=)

One important type of basic sentence is an assertion of *identity*. Propositions of this sort say that two names go with the same object, or that they're equivalent to one another. Sentences such as "Ruth is the boss," "I am the one who called," and "Istanbul is the same city as Constantinople" constitute statements of identity.

In formal logic, we can think of identity as a two-place relation. Because of its special importance, we symbolize it with an equals sign (=) instead of a predicate letter. The identity symbol goes between the two names that we want to identify, and parentheses go around the three symbols. Using this notation, we could translate "I am Jack" to $(i = j)$, or "Science is my favorite subject" to $(s = f)$.

When we identify two nouns with each other, we can interchange them without affecting the truth value of the whole. If we know that "Jack is Mr. Doe," then we can replace "Jack" with "Mr. Doe" in any sentence where we find it without changing its truth value. All the same predicates and logical relationships apply to the two names, because both names refer to the same man!

Building More Complex Formulas

Once we have a collection of complete predicate sentences, we can modify them and show their logical relationships. In order to do this, let's carry over all of the operators established for the propositional calculus in Chap. 2, along with their symbols, as follows:

- Negation ($\neg$)
- Conjunction ($\wedge$)
- Disjunction ($\vee$)
- Implication ($\rightarrow$)
- Biconditional ($\leftrightarrow$)

All the rules concerning the order of operations in propositional logic, including the use of parentheses, apply to predicate logic as well. When we start working out predicate proofs, therefore, we'll be able to use all of the propositional inference rules for the operators.

PROBLEM 3-2

Translate the following sentences into formal predicate propositions using predicate letters, names, and the logical operators:

- **You will be late, but I will not be late. (Note that "but" means the same thing as "and" in this context.)**
- **Jill and Roger are siblings. (Symbolize both "Jill" and "Roger" as proper names.)**
- **If Jack isn't walking, then he must be jumping.**
- **My grandfather is the person wearing orange. (Use two proper names.)**
- **My dog is well-trained and obedient, or she is not well-trained and unruly.**

SOLUTION

• **You (*u*) will be late (L), but I (*i*) will not be late.**	$Lu \wedge \neg Li$
• **Jill (*l*) and Roger (*r*) are siblings (S).**	Slr
• **If Jack (*j*) isn't walking (W), then he must be jumping (J).**	$\neg Wj \rightarrow Jj$
• **My grandfather (*g*) is the person wearing orange (*o*).**	$(g = o)$
• **My dog (*d*) is well-trained (B) and obedient (V), or she is not well-trained and unruly (U).**	$(Bd \wedge Vd) \vee (\neg Bd \wedge Ud)$

Quantified Statements

Using predicate letters, names, and logical operators, we can carry out all the same sorts of proofs with predicate sentences that we can do in the propositional calculus. We must go further than that, however, to take advantage of the internal structure shown by predicate propositions. To argue effectively about applying predicates to names, we can "quantify" our statements so we can talk about sets of objects like "everything that has the property F," "no two things that are in relation G to each other," or "some of the things that are both F and H." This methodology allows us to prove sequents about predicates and names in general.

Some versus All

Suppose you tell your friend that "Some dogs are terriers." You have two predicates to consider: the property of being a dog (call it D) and the property of being a terrier (T). To interpret this statement as a logical proposition, think of the evidence you need to know its truth value. A single example of an object that's both a dog and a terrier will demonstrate the validity of your statement. Therefore, the statement "Some dogs are terriers" is equivalent to "There's at least one thing that is both a dog and a terrier" or "Something exists that is both a dog and a terrier." (In common speech, "some" often means "more than one," but for our purposes it means "at least one.") Using a singular variable, x, to represent "a thing" or "something" in the formula, you can say that there exists an x that makes the conjunction $\mathrm{D}x \wedge \mathrm{T}x$ true.

Now imagine that your friend says "All dogs are mammals." Once again, you have two predicate properties. You might symbolize the property of being a mammal with the letter M. The logical interpretation of your friend's statement transforms it into a conditional proposition: "It is true of everything that if it is a dog, then it is a mammal" or "If something is a dog, then it is a mammal." These two statements have equivalent meaning, even though you might find that fact difficult to comprehend at first thought. (The statements both hold true under exactly the same conditions; in order to prove either statement false, we'd have to find a dog that's not a mammal.) Using an individual variable to translate the word "something," you can say that for anything whatsoever (represented by the variable x), $\mathrm{D}x \rightarrow \mathrm{M}x$.

The Existential Quantifier (∃)

When we translated the sentence "Some dogs are terriers" into symbols above, we left one part in ordinary language: "There exists an x." We can't leave this part out! The formal predicate language must adhere to strict rules. We call the "there is" or "there exists" portion of the proposition the *existential quantifier*, and its symbol looks like a backward capital letter E (∃).

We can read the existential quantifier out loud as "There is..." or "There exists..." or "For some..." We always follow it with a variable, and then place the character sequence in parentheses. For example, $(\exists x)$ means "There is an x" or "There exists an x" or "For some x." In a situation of this sort, we "quantify over" the variable that accompanies the ∃ symbol. For example, we might place $(\exists x)$ at the beginning of a formula, and then follow that quantifier with a proposition built up from predicate sentences that include the variable x. When fully expressed by symbols, the sentence "Some dogs are terriers" translates symbolically to

$$(\exists x)\ \mathrm{D}x \wedge \mathrm{T}x$$

If we read the above formula out loud, we say "There exists an x such that Dx and Mx." Fully translating, and letting D represent "is a dog" and T represent "is a terrier," we get "There exists an object x such that x is a dog and x is a terrier." This statement constitutes an example of an *existentially quantified proposition*. Some texts might call it an *existential statement* or a *particular sentence*.

The Universal Quantifier (∀)

We also need a symbol to stand for "For all..." or "For every..." (as in the formula for "All dogs are mammals" above). We call this symbol the *universal quantifier*. It looks like an upside-down capital letter A (∀). As with the existential quantifier, we need to "quantify over" a variable, and we use the same general syntax. We write the variable symbol immediately after the quantifier, enclosing both symbols in parentheses, usually at the beginning of a proposition. [Some texts place a variable alone within parenthesis to represent universal quantification, for example (x) rather than $(\forall x)$.] According to the convention that we will use, the plain-language sentence "All dogs are mammals" translates to the symbol sequence

$$(\forall x)\ \mathrm{D}x \rightarrow \mathrm{M}x$$

Reading the above formula out loud, we say "For all x, if Dx, then Mx." Fully translating, we obtain the sentence "For all objects x, if x is a dog, then x is a mammal." This statement demonstrates an example of a *universally quantified proposition* (also known as a *universal statement* or *general sentence*).

The Universe

In order to know that a simple *existential* statement holds true, you need to produce only one example of the property or properties attributed to the variable. The example might be "definite," using a proper noun, or "arbitrary," with an arbitrary name. If you want to ensure that a *universal* statement holds true, however, you must demonstrate that it holds true for every possible example. A single counterexample will invalidate it, no matter how many examples make it seem true. In fact, a single counterexample will invalidate a statement for which *infinitely many* other examples work!

What counts as "every possible example" depends on the context. If we discuss siblings, then "every possible example" might constitute the set of all humans (a large but finite set). If we talk about mathematics, then "every possible example" might constitute the set of all real numbers (an infinite set). We call such a set the *universe of discourse* or simply the *universe*; some texts call it the *domain of discourse*. In many cases, we can take for granted the sorts of objects to which a predicate can apply; in other situations we must define the universe explicitly (especially when using quantifiers).

Thinking about and clearly defining the universe can help make sense of some quantified statements. We can imagine the existential quantifier as a long disjunction. For example, the formula

$$(\exists x)\ Fx$$

is like the long disjunction

$$Fm \vee Fn \vee Fo \ldots$$

where the set $\{m, n, o, \ldots\}$, the universe, contains every object—but only those objects—to which the predicate might apply. If any of the disjuncts is true, then the predicate is true of something; a single example is all we need to demonstrate the validity of the statement. We can refer to the corresponding arbitrary statement Fa as the *typical disjunct* or *representative instance* of $(\exists x)\ Fx$, because it can stand for any element of the set of statements $\{Fm, Fn, Fo, \ldots\}$.

Following along similar lines, we can imagine the universal quantifier as a long conjunction; the formula

$$(\forall x)\ Fx$$

is like the long conjunction

$$Fm \wedge Fn \wedge Fo \wedge \ldots$$

where the set $\{m, n, o, \ldots\}$, the universe, contains every object—but only those objects—to which the predicate might apply. The whole conjunction holds true if and only if every conjunct is true; if only one of the conjuncts is false, then the whole conjunction is false. We can refer to the arbitrary statement Fa as the *typical conjunct* or representative instance of $(\forall x)\ Fx$, because it can stand for any element of the set $\{Fm, Fn, Fo, \ldots\}$.

TIP ***The universe may contain a finite number of things, but in many instances we'll encounter universes that contain infinitely many elements (the set of all positive whole numbers or the set of all real numbers, for instance). In a situation like that, we can't write out a quantifier's equivalent conjunction or disjunction completely, but we can nevertheless use this device as a way of thinking about the truth conditions of quantified propositions. In this book, we'll always assume that the universe contains at least one element. We won't get into the esoteric business of ruminating over empty universes!***

Multiple Quantifiers

A single sentence can have more than one quantifier, each applying to a different variable. For instance, you might say, "Some people have a sibling." To express this sentence symbolically using the sibling relation S, we need two existential quantifiers (one for "some people" and one for their unnamed siblings), producing the formula

$$(\exists x)(\exists y)\ Sxy$$

When all the quantifiers in a multiple-quantifier formula are of the same type (universal or existential, as in this example), then we can list the quantifier once, with all the applicable variables listed immediately to its right, obtaining

$$(\exists x, y)\ Sxy$$

We would read this out loud as "There exist an x and a y such that Sxy."

When using multiple variables and quantifiers, we must always assign a unique variable to each constant in a formula. In the foregoing situation, we wouldn't want to write a formula such as

$$(\exists z)\ Szz$$

which, translated literally, means "There exists a z who is his or her own sibling."

If we want to construct a proposition that requires both a universal quantifier and an existential quantifier, we must preserve the orders of the quantifiers and variable assignments. Otherwise, we might commit the "fallacy of every and all," an example of which is

$$(\forall y)(\exists x)\ Rxy$$

If we let R mean "is an ancestor of," then this formula translates to "Everyone has somebody else as an ancestor." It's obviously not equivalent to

$$(\exists x)(\forall y)\ Rxy$$

which translates to "There is someone who is an ancestor of everybody else."

PROBLEM 3-3

Translate the following English sentences into formal predicate propositions using quantifiers and variables (use *x*) in addition to the other symbols you've learned:

- **All dogs are well-trained.**
- **Somebody is Jack's sibling. (Use a two-part relation.)**
- **Some Americans have colds.**
- **For any pair of siblings, one is older than the other.**
- **Everybody was late.**

SOLUTION

You can assign predicates and variables, and then break down the sentences into formal predicate propositions as follows:

• **All dogs (D) are well-trained (B).**	$(\forall x)\ Dx \rightarrow Bx$
• **Somebody is Jack's (*j*) sibling (S).**	$(\exists x)\ Sxj$
• **Some Americans (A) have colds (C).**	$(\exists x)\ Ax \wedge Cx$
• **For any pair of siblings (S), one is older(O) than the other.**	$(\forall x, y)\ Sxy \rightarrow Oxy \vee Oyx$
• **Everybody was late (L).**	$(\forall x)\ Lx$

These are the most straightforward, recognizable ways to symbolize these sentences. We can make up plenty of other, logically equivalent formulas that hold true under the same conditions.

Translating the Quantifiers

Subtle distinctions occur when we translate the existential quantifier to the universal, or vice versa. When we want to assert a combination of properties, the existential quantifier employs conjunction, while we express universal statements using the conditional (if/then) operator. For example, "Some Americans have colds" translates to

$$(\exists x)\ \mathrm{A}x \wedge \mathrm{C}x$$

but "All Americans have colds" translates to

$$(\forall x)\ \mathrm{A}x \rightarrow \mathrm{C}x$$

We must use the applicable connecting operator ($\wedge$ with $\exists$ and $\rightarrow$ with $\forall$), or we might lose the meaning in translation. For example, if we write

$$(\forall x)\ \mathrm{A}x \wedge \mathrm{C}x$$

using the above assignments, we say, in effect, that "Everyone is an American with a cold." That wouldn't hold true even if everybody in the world had a cold!

Laws of Quantifier Transformation

We can transform any quantified statement from universal to existential or vice versa using the negation operator. For instance, the universal sentence "Everybody was late," symbolized as

$$(\forall x)\ \mathrm{L}x$$

means the same things as the existential sentence "There was nobody who was not late," which we symbolize as

$$\neg(\exists x)\ \neg\mathrm{L}x$$

The negation of the first statement, "Not everyone was late," symbolized as

$$\neg(\forall x)\ \mathrm{L}x$$

holds true under the same conditions as the negation of the second statement, "Someone was not late," symbolized as

$$(\exists x)\ \neg\mathrm{L}x$$

We can state two general equivalences of this kind, called the *laws of quantifier transformation*. (These rules might remind you of DeMorgan's laws, which transform between conjunction and disjunction using negation.) They show that the quantifiers are *interderivable*, meaning that we can reformulate any quantified statement in terms of a single quantifier (either universal or existential). We express the laws of quantifier transformation as

$$(\exists x)\ Fx \dashv\vdash \neg(\forall x)\ \neg Fx$$

and

$$(\forall x)\ Fx \dashv\vdash \neg(\exists x)\ \neg Fx$$

Both propositions in either transformation have the same truth value, so their negations must have the same truth value, too. By negating both sides of the equivalence sequents and eliminating double negation, we can see the laws in the forms

$$\neg(\exists x)\ Fx \dashv\vdash (\forall x)\ \neg Fx$$

and

$$\neg(\forall x)\ Fx \dashv\vdash (\exists x)\ \neg Fx$$

Logical Relationships: The Square of Opposition

The logical relationships between quantified statements can be summed up in a diagram called a *square of opposition* (Fig. 3-1). Propositions located at

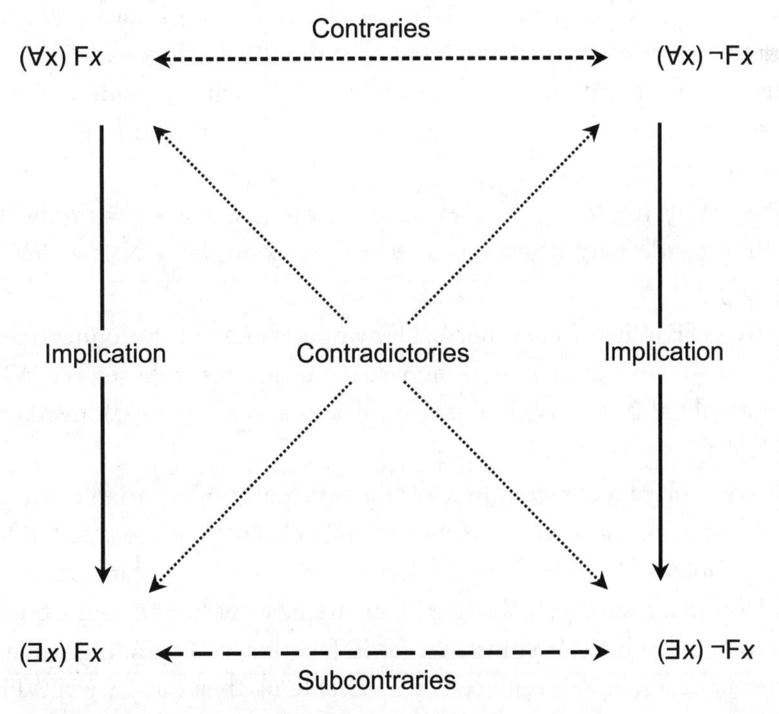

FIGURE 3-1 • The square of opposition for logical quantifiers.

opposite corners contradict (they negate each other). The universal propositions on top imply the existential propositions directly below them; if a property applies to everything, then it applies to something. We call the universal statements on the top of the square *contraries* (they never hold true at the same time, but might both be false) and the existential statements on the bottom of the square *subcontraries* (they are never both false, but both may hold true).

Well-Formed Formulas

If we want to ensure that our formulas express genuine propositions in predicate logic, we need a set of rules against which we can test them. If a formula follows the rules, we call it a *well-formed formula* (WFF). The following rules define the sorts of formulas that qualify as WFFs in predicate logic.

- Any PF (as defined in Chap. 2) constitutes a WFF.
- Any subject/predicate formula in which a predicate letter is followed by a finite number of terms, such as Ak, constitutes an atomic sentence, and all atomic sentences are WFFs. For example, if A is a predicate and k is a constant, then Ak is a WFF; if k_1, k_2, k_3,...k_n are constants, then Ak_1k_2...k_n is a WFF. (For a propositional variable, $n = 0$.) Identity is a special predicate relation, and the identity symbol is analogous to a predicate letter. An identity statement like $(k_1 = k_2)$, written inside parentheses, constitutes an atomic sentence. For example, if k_1 and k_2 are constants, then $(k_1 = k_2)$ is a WFF.
- Any WFF immediately preceded by the symbol ¬, with the whole thing inside parentheses, is a WFF. For example, if X is a WFF, then (¬X) is a WFF.
- Any WFF joined to another WFF by one of the binary connectives ∧, ∨, →, or ↔, with the whole thing written inside parentheses, is a WFF. For example, if X is a WFF, Y is a WFF, and # is a binary connective, then (X # Y) is a WFF.
- If we replace a constant in a WFF consistently by a variable, this action constitutes a *substitution instance* of the WFF. If we place a quantifier and a variable in parentheses to the left of a substitution instance, the result is a WFF. For example, if X is a WFF including a constant k, @ is a quantifier, and Y is a substitution instance of X (the formula resulting when every instance of k in X is replaced by a variable v), then (@v) Y is a WFF.

- Any formula that does not satisfy one or more of the five conditions listed above does not constitute a WFF.

You can drop some of the parentheses from your formulas according to the conventional order of operations. We did that in some of the examples above; for example, the rigorous form of

$$(\mathrm{B}d \wedge \mathrm{V}d) \vee (\neg \mathrm{B}d \wedge \mathrm{U}d)$$

would be written as

$$\{(\mathrm{B}d \wedge \mathrm{V}d) \vee [(\neg \mathrm{B}d) \wedge \mathrm{U}d]\}$$

with two extra sets of grouping symbols. You can drop the extra grouping symbols to make WFFs easier to read, but you can substitute them back in at any time to satisfy the strict definition of a WFF above, or whenever doing so makes a formula easier for you to understand.

TIP ***A statement can form a legitimate WFF even if it doesn't hold true. Statements whose truth value remains unknown, and even statements that are obviously false, can constitute perfectly good WFFs! They only need to follow the foregoing rules concerning the arrangements of the symbols.***

Properties of Two-Part Relations

Two-part predicates have a set of logical categories all their own, called the *properties of relations*. Let's define the most important general properties.

Symmetry

A relation is *symmetric* if it makes the following proposition true whenever we substitute it for F in the formula

$$(\forall x, y)\ \mathrm{F}xy \rightarrow \mathrm{F}yx$$

This rule tells us that we can always transpose or rearrange the order of the variables or constants without changing the truth value. The property of "being a sibling," for example, is a symmetric relation. If Jill is Roger's sibling, then

Roger is Jill's sibling. Identity is also symmetric: if $x = y$, then $y = x$. Because we state the rule universally, one counterexample will demonstrate that a relation is not symmetric.

Asymmetry

If a relation holds true in one "direction" but not the other, we call that relation *asymmetric*. Using the same symbol convention as we did when we stated the criterion for symmetry, we can express the criterion for asymmetry as

$$(\forall x, y)\ Fxy \rightarrow \neg Fyx$$

The property of "being older than someone" is asymmetric. If Jill is older than Roger, then Roger is not older than Jill. If we can find one case where the order of names doesn't affect the proposition truth value of the proposition F*xy*, then we've proved that F isn't asymmetric.

Antisymmetry

Sometimes a predicate follows the rule of asymmetry except when we use the same name twice. Consider the mathematical relations "less than or equal to" (≤) and "greater than or equal to" (≥). The asymmetry rule holds true of these relations for most numbers we can assign to x and y, but not when we choose the same number for both variables. We call this type of relation *antisymmetric,* and we can express the criterion as

$$(\forall x, y)\ Fxy \wedge Fyx \rightarrow (x = y)$$

Nonsymmetry

A relation that's neither symmetric nor antisymmetric (and therefore not asymmetric, either) is known as *nonsymmetric*. Now we've covered everything!

TIP ***We can always put any given relation in the predicate calculus into one or more of the above-defined four categories of symmetry.***

Reflexivity

If a predicate relation holds between any name and itself, then we call that relation *reflexive*. For example, the following relation exhibits *reflexivity*:

$$(\forall x)\ Fxx$$

We need not work very hard to come up with specific cases of reflexivity. Everything is identical to itself, every woman is the same age as herself, and so on. If we can come up with a single example of a constant where the predicate does not hold true in this way, then the relation is not reflexive.

Irreflexivity

If a relation cannot hold between any constant and itself, like the property of being older than someone, then we call that relation *irreflexive*. We can define this property by writing the relation

$$(\forall x)\ \neg Fxx$$

Nonreflexivity

If some substitution instances make Fxx true and others make $\neg Fxx$ true, then the relation F does not meet the definition for reflexivity or irreflexivity. In a case of this sort, we categorize the relation as *nonreflexive*.

TIP ***Any given relation will satisfy only one of the foregoing three criteria; it must be either reflexive, irreflexive, or nonreflexive.***

Transitivity

Suppose we have a relation that holds true between x and y and also between y and z. We call that relation *transitive* if it also holds true between x and z. We can define *transitivity* in symbolic terms as

$$(\forall x, y, z)\ Fxy \wedge Fyz \rightarrow Fxz$$

The relation of "being older" provides us with an example of transitivity. If Ruth is older than Jill and Jill is older than Roger, then we know that Ruth must be older than Roger. Identity also constitutes a transitive relation. Can you think of others?

Intratransitivity

Any case that contradicts the universal proposition above indicates that transitivity does not hold for the relation in question; when no examples show this pattern, we call the relation *intransitive*. We can define *intransitivity* in symbolic terms as

$$(\forall x, y, z)\ Fxy \wedge Fyz \rightarrow \neg Fxz$$

For instance, if Tuesday is the day immediately before Wednesday, and Wednesday is the day immediately before Thursday, then Tuesday can't be the day immediately before Thursday.

Nontransitivity

If a relation meets neither of the preceding two criteria, then we call it *nontransitive*. For this type of relation, we could produce counterexamples to the universal statements that express transitivity and intransitivity.

TIP ***Any given two-part relation will satisfy* only one *of the foregoing three criteria; it must be either transitive, intratransitive, or nontransitive.***

Equivalence Relations

We can always describe a two-part predicate relation, no matter how complicated or esoteric, in terms of the symmetry, reflexivity, and transitivity criteria and their variants. If a relation meets the criteria for symmetry, reflexivity, and transitivity "all at the same time," then we call it an *equivalence relation*. We have established that identity, an especially important relation in logic, constitutes an equivalence relation.

Predicate Proofs

With the symbols of the predicate calculus we can make rigorous arguments that take advantage of sentence structure. All of the proof methods, inference rules, and laws that we learned in propositional logic still apply, but predicate proofs require us to adopt some new inference rules.

Existential Introduction (∃I)

A single instance of a predicate applied to a constant can serve as a premise to derive an existential statement about a variable. We replace the constant consistently using the same variable that goes next to the quantifier. (We should, of course, choose a new variable letter that we haven't used elsewhere in the proof.) We call this principle the *rule of existential introduction*; we can symbolize it as ∃I. To implement it during the course of a proof, we cite the atomic sentence in the reference column and carry over its assumptions, as shown in Table 3-1.

TABLE 3-1 An example of existential introduction.

Line number	Proposition	Reference	Rule	Assumption
1	F*m*		A	1
2	(∃*x*) F*x*	1	∃I	1

Existential Elimination (∃E)

We can derive conclusions using an existential statement as a premise by invoking the *rule of existential elimination*. If we can prove the desired conclusion using an "arbitrary example" of a property (that is, an atomic sentence that includes an arbitrary name) assumed as a premise, then we can derive the same conclusion from a statement that says something with that property exists. The assumed premise should look like the existential proposition, but with the quantifier dropped and the arbitrary name consistently replacing the existentially quantified variable.

To take advantage of this rule, we need to reference three lines: the arbitrary sentence used as a premise, the statement proved from it, and the existential premise. From the assumptions listed with the first statement of the conclusion, we drop the assumption corresponding to the arbitrary premise and replace it with the assumption(s) of the existential statement. (This procedure may remind you of disjunction elimination; we derive a conclusion and then "swap out" the assumptions.) In the rule column, we write ∃E.

The arbitrary name constitutes a temporary device in this kind of derivation, used as a typical disjunct for the existential proposition. We must treat it carefully in order to avoid generating a fallacious argument. In the final step, where

TABLE 3-2 An example of existential elimination.

Line number	Proposition	Reference	Rule	Assumption
1	$(\exists x)$ Fx		A	1
2	F$a \to$ Ga	(already derived)		*
3	Fa		A	3
4	Ga	2, 3	MP	3, *
5	$(\exists x)$ Gx	4	∃I	3, *
6	$(\exists x)$ Gx	1, 3, 5	∃E	1, *

we apply ∃E, we drop the assumption involving the arbitrary name. To ensure a valid deduction, that arbitrary name should not appear in the conclusion itself or in any of the remaining assumptions.

The example in Table 3-2 starts off "in the middle," as if we had already proved the arbitrary sentence at line 2, F$a \to$ Ga, on the basis of some unknown assumptions signified by an asterisk (∗). We can assume that no proposition in ∗ includes the arbitrary name a. Doing that keeps the proof simple without violating the restriction on assuming a proposition with arbitrary name.

Universal Introduction (∀I)

In an atomic proposition with an arbitrary name such as Fa where the arbitrary name stands for "anything"; we can understand the statement to mean "Take anything you like; that thing has property F." If property F applies to anything that we might happen to pick out, then we can conclude that F must apply to everything. We call this principle the *rule of universal introduction*, abbreviated ∀I. When employing this rule, the arbitrary statement is the only premise referenced, and all of its assumptions are retained.

Whenever we invoke ∀I, the arbitrary name should not appear in any of the assumptions cited alongside the conclusion. The conclusion itself constitutes a universally quantified proposition with every appearance of the arbitrary name replaced by the quantified variable.

Table 3-3 begins with the typical conjunct Fa, as if we had already derived it from a set of assumptions signified by an asterisk (∗). We can assume that no proposition in ∗ includes the arbitrary name a. Making this assumption keeps the proof as simple as possible without violating the restriction on assuming a proposition with an arbitrary name.

TABLE 3-3 An example of universal introduction.

Line number	Proposition	Reference	Rule	Assumption
1	F*a*	(already derived)		*
2	(∀*x*) F*x*	1	∀I	*

Universal Elimination (∀E)

A universal statement holds true only if every single atomic sentence we can produce from it is true, so we can use the general statement as a premise to derive one of its instances. To take advantage of the *rule of universal elimination*, we reference the universal as a premise, carry over its assumptions, and fill in any constant we like, consistently substituted in place of the universally quantified variable. In the rule column, we write ∀E as shown in Table 3-4.

TABLE 3-4 An example of universal elimination.

Line number	Proposition	Reference	Rule	Assumption
1	(∀*x*) F*x*		A	1
2	F*m*	1	∀E	1

Identity Introduction (=I)

When we want to bring a statement of identity into a proof, we can use the *rule of identity introduction*. This inference rule derives from the trivial reasoning that any name is self-identical (everything is itself!). We can always substitute a name for itself without changing the truth value of a proposition, because the resulting formula looks exactly the same as it did before. Using identity introduction, we can introduce a proposition of the form $(m = m)$ at any point in a proof without having to cite any premises or assumptions whatsoever. We write =I in the rule column to indicate the use of this rule, as shown in Table 3-5.

TABLE 3-5 An example of identity introduction.

Line number	Proposition	Reference	Rule	Assumption
1	$(m = m)$		=I	

TABLE 3-6 An example of identity elimination.

Line number	Proposition	Reference	Rule	Assumption
1	$(m = n)$		A	1
2	Fm		A	2
3	Fn	1, 2	=E	1, 2

Identity Elimination (=E)

If two names are identical, then we can substitute one for another. This principle is the basis for the *rule of identity elimination*, symbolized =E. To use the rule, we reference two lines as premises: an identity statement and another proposition including one (or both) of the identified names. Then we can derive a substitution instance of the target proposition, "swapping out" one name for another. We don't have to substitute all instances consistently; we can choose at will between the names in each instance where one of them appears, and finally combine the assumptions of the two premises. Table 3-6 illustrates how =E works.

Conventions for Predicate Proofs

In predicate proofs, we can quantify the assumptions and conclusions (propositions appearing in the proved sequent) to keep arbitrary constants out of sequents. As the restrictions on universal introduction and existential elimination suggest, we should strive to use arbitrary names only on a temporary basis. We can "extract" them from quantified statements using the quantifier inference rules, manipulate the resulting arbitrary atomic sentences with propositional transformations, and then "requantify," substituting all the arbitrary constants with quantified variables in the conclusion.

PROBLEM 3-4

Use the inference rules to prove the sequent

$$(\exists x)\ Fx \wedge Gx \vdash (\exists x)\ Fx \wedge (\exists x)\ Gx$$

SOLUTION

See Table 3-7.

TABLE 3-7 Proof that $(\exists x)\ Fx \land Gx \vdash (\exists x)\ Fx \land (\exists x)\ Gx$.

Line number	Proposition	Reference	Rule	Assumption
1	$(\exists x)\ Fx \land Gx$		A	1
2	$Fa \land Ga$		A	2
3	Fa	2	∧E	2
4	$(\exists x)\ Fx$	3	∃I	2
5	Ga	2	∧E	2
6	$(\exists x)\ Gx$	5	∃I	2
7	$(\exists x)\ Fx \land (\exists x)\ Gx$	4, 6	∧I	2
8	$(\exists x)\ Fx \land (\exists x)\ Gx$	1, 2, 7	∃E	1

PROBLEM 3-5

Use the inference rules to prove the sequent

$$(\forall x)\ Fx \rightarrow Gx, \neg Gm \vdash (\exists x)\ \neg Fx$$

SOLUTION

See Table 3-8.

PROBLEM 3-6

Use the inference rules to prove the sequent

$$(\exists x)\ Fx \land \neg Gx, (\forall x)\ Fx \rightarrow Hx \vdash (\exists x)\ Hx \land \neg Gx$$

SOLUTION

See Table 3-9.

TABLE 3-8 Proof that $(\forall x)\ Fx \rightarrow Gx, \neg Gm \vdash (\exists x)\ \neg Fx$.

Line number	Proposition	Reference	Rule	Assumption
1	$(\forall x)\ Fx \rightarrow Gx$		A	1
2	$Fm \rightarrow Gm$	1	∀E	1
3	$\neg Gm$		A	3
4	$\neg Fm$	2, 3	MT	1, 3
5	$(\exists x)\ \neg Fx$	4	∃I	1, 3

TABLE 3-9 Proof that $(\exists x)\ Fx \wedge \neg Gx, (\forall x)\ Fx \rightarrow Hx \vdash (\exists x)\ Hx \wedge \neg Gx$.

Line number	Proposition	Reference	Rule	Assumption
1	$(\exists x)\ Fx \wedge \neg Gx$		A	1
2	$(\forall x)\ Fx \rightarrow Hx$		A	2
3	$Fa \rightarrow Ha$	2	∀E	2
4	$Fa \wedge \neg Ga$		A	4
5	Fa	4	∧E	4
6	Ha	3, 5	MP	2, 4
7	$\neg Ga$	4	∧E	4
8	$Ha \wedge \neg Ga$	6, 7	∧I	2, 4
9	$(\exists x)\ Hx \wedge \neg Gx$	8	∃I	2, 4
10	$(\exists x)\ Hx \wedge \neg Gx$	1, 4, 9	∃E	1, 2

PROBLEM 3-7

Use the inference rules to prove the sequent

$$(\forall x)\ Fx \wedge (\forall x)\ Gx \vdash (\forall x)\ Fx \wedge Gx$$

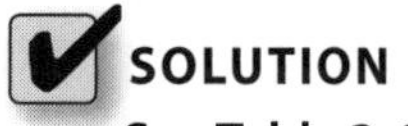

SOLUTION

See Table 3-10.

TABLE 3-10 Proof that $(\forall x)\ Fx \wedge (\forall x)\ Gx \vdash (\forall x)\ Fx \wedge Gx$.

Line number	Proposition	Reference	Rule	Assumption
1	$(\forall x)\ Fx \wedge (\forall x)\ Gx$		A	1
2	$(\forall x)\ Fx$	1	∧E	1
3	$(\forall x)\ Gx$	1	∧E	1
4	Fa	2	∀E	1
5	Ga	3	∀E	1
6	$Fa \wedge Ga$	4, 5	∧I	1
7	$(\forall x)\ Fx \wedge Gx$	6	∀I	1

PROBLEM 3-8

Use the inference rules to prove the laws of quantifier transformation

$$(\exists x)\ Fx \dashv\vdash \neg(\forall x)\ \neg Fx$$

and

$$(\forall x)\ Fx \dashv\vdash \neg(\exists x)\ \neg Fx$$

SOLUTION

Both laws constitute interderivability sequents, so we must execute a total of four proofs to back them up. Tables 3-11 through 3-14 show the derivations.

TABLE 3-11 Proof that $(\exists x)\ Fx \vdash \neg(\forall x)\ \neg Fx$.

Line number	Proposition	Reference	Rule	Assumption
1	$(\exists x)\ Fx$		A	1
2	$(\forall x)\ \neg Fx$		A	2
3	$\neg Fa$	2	∀E	2
4	Fa		A	4
5	$Fa \wedge \neg Fa$	3, 4	∧I	2, 4
6	$\neg(\forall x)\ \neg Fx$	2, 5	RAA	4
7	$\neg(\forall x)\ \neg Fx$	1, 4, 6	∃E	1

TABLE 3-12 Proof that $\neg(\forall x)\ \neg Fx \vdash (\exists x)\ Fx$.

Line number	Proposition	Reference	Rule	Assumption
1	$\neg(\forall x)\ \neg Fx$		A	1
2	$\neg(\exists x)\ Fx$		A	2
3	Fa		A	3
4	$(\exists x)\ Fx$	3	∃I	3
5	$(\exists x)\ Fx \wedge \neg(\exists x)\ Fx$	2, 4	∧I	2, 3
6	$\neg Fa$	3, 5	RAA	2
7	$(\forall x)\ \neg Fx$	6	∀I	2
8	$(\forall x)\ \neg Fx \wedge \neg(\forall x)\ \neg Fx$	1, 7	∧I	1, 2
9	$\neg\neg(\exists x)\ Fx$	2, 8	RAA	1
10	$(\exists x)\ Fx$	9	DN	1

TABLE 3-13 Proof that $(\forall x)\ Fx \vdash \neg(\exists x)\ \neg Fx$.

Line number	Proposition	Reference	Rule	Assumption
1	$(\forall x)\ Fx$		A	1
2	$(\exists x)\ \neg Fx$		A	2
3	$\neg Fa$		A	3
4	Fa	1	∀E	1
5	$Fa \wedge \neg Fa$	3, 4	∧I	1, 3
6	$\neg(\forall x)\ Fx$	1, 5	RAA	3
7	$(\forall x)\ Fx \wedge \neg(\forall x)\ Fx$	1, 6	∧I	1, 3
8	$(\forall x)\ Fx \wedge \neg(\forall x)\ Fx$	2, 3, 7	∃E	1, 2
9	$\neg(\exists x)\ \neg Fx$	2, 8	RAA	1

TABLE 3-14 Proof that $\neg(\exists x)\ \neg Fx \vdash (\forall x)\ Fx$.

Line number	Proposition	Reference	Rule	Assumption
1	$\neg(\exists x)\ \neg Fx$		A	1
2	$\neg Fa$		A	2
3	$(\exists x)\ \neg Fx$	2	∃I	2
4	$(\exists x)\ Fx \wedge \neg(\exists x)\ Fx$	1, 3	∧I	1, 2
5	$\neg\neg Fa$	2, 4	RAA	1
6	Fa	5	DN	1
7	$(\forall x)\ Fx$	6	∀I	1

Syllogisms

The term *syllogism* refers to a well-known argument form in three lines with an alternate symbolism. We can use a syllogism as an abbreviated way to do certain predicate proofs. *Syllogistic* arguments constitute a subset of all possible predicate arguments. Not all predicate proofs can be expressed as syllogisms, and not all WFFs of the predicate calculus can appear in a syllogism, but all syllogisms can be expressed in the formal predicate system (if we follow some restrictions) and proved according to the conventional rules.

A Sample Syllogism

The following proof, expressed in everyday language, gives us a simple example of a syllogism:

All dogs are mammals.

All terriers are dogs.

Therefore, all terriers are mammals.

This argument is a *traditional syllogism* (also called a *categorical syllogism*), which means that it has three lines; each of the three propositions is quantified and contains two predicates.

Allowed Propositions

We will find four types of proposition used in a traditional syllogism. We call such propositions *categorical statements*, and they have their own symbolism.

A *universal affirmative* proposition (abbreviated **A**) has the form

$$(\forall x)\ \mathrm{F}x \rightarrow \mathrm{G}x$$

like the statement "All dogs are mammals." In syllogistic notation, we denote this form as

$$\mathbf{A}(\mathrm{F},\mathrm{G})$$

A *universal negative* proposition (**E**) has the form

$$(\forall x)\ \mathrm{F}x \rightarrow \neg\mathrm{G}x$$

like the statement "Anything that is an insect cannot be a mammal." In syllogistic notation, we denote this form as

$$\mathbf{E}(\mathrm{F},\mathrm{G})$$

A *particular affirmative* proposition (**I**), has the form

$$(\exists x)\ \mathrm{F}x \wedge \mathrm{G}x$$

like the statement "There is at least one dog that is a mammal." In syllogistic notation, we denote this form as

$$I(F,G)$$

A *particular negative* proposition (O) has the form

$$(\exists x)\ Fx \wedge \neg Gx$$

like the statement "Some fish are not mammals." In syllogistic notation, we denote this form as

$$O(F,G)$$

The WFF equivalents of the four proposition types are all quantified over a single variable.

The three-line argument stated a while ago (concerning dogs and mammals), rewritten in syllogistic notation, consists entirely of universal affirmatives. The original syllogism (in words) appears on the left below, while the symbolic version appears on the right. Let D = dog, M = mammal, and T = terrier. Then we can write

All dogs are mammals.	A(D,M)
All terriers are dogs.	A(T,D)
All terriers are mammals.	A(T,M)

Terms and Arguments

In a syllogism, we call the letters representing predicates the *terms*. (We must take care to avoid confusing syllogistic terms with proper or arbitrary names, also sometimes referred to as "terms" when they don't appear in syllogisms.) We can call the first term the *subject term* and the second term the *predicate term*, corresponding to the "subject/predicate" order that we commonly use in everyday speech.

Every complete syllogism involves three terms. The *major term* appears in the first premise and the conclusion; in the above example, M constitutes the major term. The *minor term* is used in the second premise and the conclusion, like T in the above sample argument. The *middle term* always occurs in both premises, but not in the conclusion; D fills this role in the above argument. On the basis of the terms they include, we call the premises the *major premise* (the first) and the *minor premise* (the second). Therefore, we can generalize the above argument as follows:

A(D,M)	Major premise	(middle and major term)
A(T,D)	Minor premise	(minor and middle term)
A(T,M)	Conclusion	(minor and major term)

These rules help us to simplify formal syllogisms, but they don't encompass every possibility. Syllogisms in everyday speech might deviate from these conventions and nevertheless produce valid arguments.

TIP ***We call a term*** **distributed** ***if it refers to everything with that property as used in the proposition. If we know the proposition type and position of the term (first or second in the statement), then the distribution appears as shown in Table 3-15 (with*** **d** ***for distributed,*** **u** ***for undistributed).***

TABLE 3-15 Distribution of terms.

A(d,u)	E(d,d)
I(u,u)	O(u,d)

Conversion Laws

We can reverse the order of the terms in the universal negative (E) and the particular affirmative (I) forms without changing the truth value. Two laws summarize these allowed reversals: the *law of the conversion of* **E** and the *law of the conversion of* ***I***. We can state them symbolically as

$$E(F,G) \leftrightarrow E(G,F)$$

and

$$I(F,G) \leftrightarrow I(G,F)$$

Classifying Syllogisms: Mood and Figure

Because the rules governing syllogistic form are strictly limited (three lines and four proposition types), only a finite number of possible argument patterns can constitute well-formed syllogisms. We can classify well-formed syllogisms according to the pattern of the propositions (called the *mood*) and the arrangement of terms within the propositions (called the *figure*).

We define the mood of a syllogism according to the three proposition types that appear along the left side of the argument; we can name them using the three letters in the order they appear. The mood of the foregoing sample argument is **AAA**. There are four possible types of proposition for each line of the syllogism, which makes for 64 possible moods ($4 \times 4 \times 4 = 4^3 = 64$).

We define the figure of a syllogism as the four possible arrangements of terms (if we ignore the proposition types). If we call the minor term F, the major term G, and the middle term H, the four figures in the foregoing syllogism are as follows:

1	2	3	4
(H,G)	(G,H)	(H,G)	(G,H)
(F,H)	(F,H)	(H,F)	(H,F)
___	___	___	___
(F,G)	(F,G)	(F,G)	(F,G)

If we know both the mood and the figure of a syllogism, then we can uniquely define the logical form of the syllogism. The actual terms used to fill in the blanks are not logically important. Any of the four figures can be combined with any of the 64 moods, giving us a total of 256 possible patterns ($4 \times 64 = 256$). We can identify any specific pattern by writing out the mood followed by the figure. The foregoing sample argument has the pattern of the first figure, so we can call it **AAA1**.

Testing Syllogistic Arguments

We can test the validity of syllogisms using any of several different techniques. Specialized diagrams known as *Venn diagrams*, adapted from set theory in pure mathematics, provide a powerful tool for this purpose. When used for syllogism-testing, a Venn diagram contains three circles, one for each term. We represent both premises on the same diagram. If the resulting picture illustrates the conclusion, then we know that we have a valid argument.

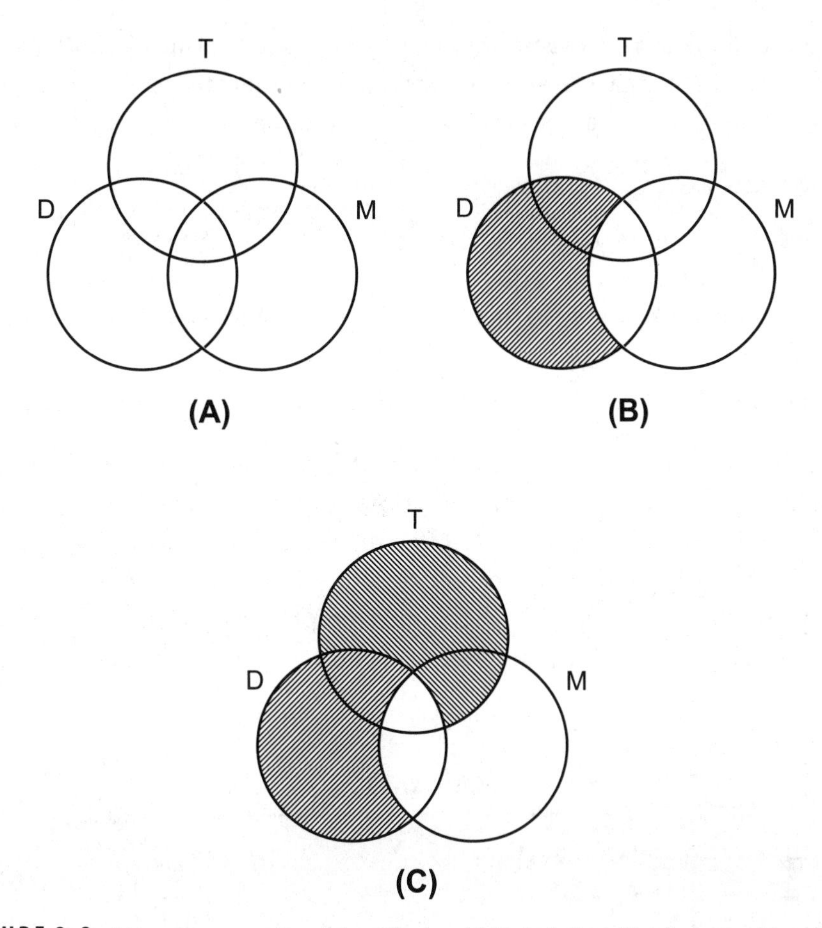

FIGURE 3-2 • A Venn interpretation of the syllogism AAA1. At A, the initial diagram, with mutually intersecting empty circles for D (dogs), M (mammals), and T (terriers). At B, the illustration for A(D,M), "All dogs are mammals." At C, we add the illustration for A(T,D), "All terriers are dogs," obtaining the conclusion A(T,M), "All terriers are mammals."

To draw the diagram for **AAA**1 as it appears in the foregoing example, we start with three white circles, intersecting as shown in Fig. 3-2A. Then we shade or hatch (to eliminate) the region within the "dogs" (D) circle that *does not* intersect with "mammals" (M) to represent the first premise, **A**(D,M), as illustrated in Fig 3-2B. Next, for **A**(T,D), we shade or hatch (to eliminate) the region within the "terriers" (T) circle that *does not* intersect with "dogs" (D) on the same picture (Fig. 3-2C). To express the conclusion of the argument, **A**(T,M), we must shade or hatch the region within the "terriers" (T) circle that *does not* intersect with the "mammals" (M) circle—but we've already done that! The premises in combination determine the conclusion, so we know that our syllogism is valid.

We can also test a syllogistic argument using a set of principles intended to avoid fallacies. A formal syllogism, constructed according to the formation rules we've learned in this chapter, is valid if it satisfies the following conditions:

- At least one of the premises is affirmative (type **A** or **I**).
- If one premise is negative (type **E** or **O**), the conclusion must also be negative.
- If a term is distributed in the conclusion, it must also be distributed in the premise.
- The middle term must be distributed in at least one of the premises.

Some logicians draw a so-called *square of opposition* that shows the logical relationships between the syllogistic proposition types (Fig. 3-3). To ensure that these logical relationships to hold true, we must make some existential assumptions. Some of these logical relationships have *existential import*, which means that they do not hold if we fill in an "empty term" where the predicate does not apply to anything in the universe of discourse. (Traditional syllogisms

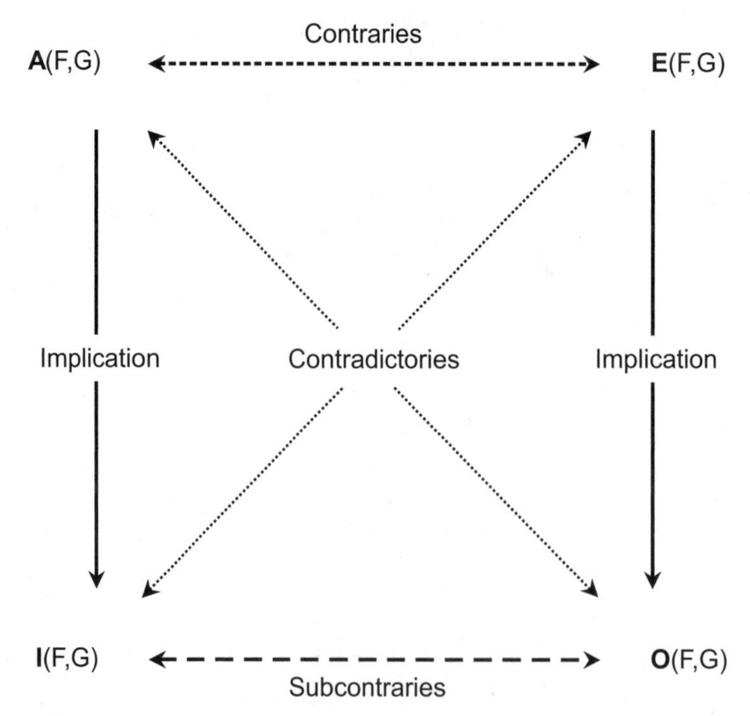

FIGURE 3-3 • The traditional square of opposition for categorical propositions.

assume that all terms *must* apply to *something*.) When looking at the diagram, we assume that no empty terms will exist. To demonstrate one of these relationships using a predicate proof, we may need to make an extra assumption of the form $(\exists x)\ Fx$.

Valid Syllogisms

Only 24 of the 256 possible syllogisms constitute valid arguments (six per figure), as shown in Table 3-16.

The valid forms of syllogism have traditional nicknames according to their mood. Our example goes by the most famous of these names, "BARBARA." The names give us a mnemonic device where the vowels spell out the mood; so, for example, BARBARA means the valid form of **AAA** (the one using the first figure, **AAA**1).

The starred forms in Table 3-16 have existential import, so each contains a "hidden" existential assumption about one of its predicates.

TABLE 3-16 Syllogisms that constitute valid arguments. Starred items have existential import.

AAA1	AEE2	AAI3*	AAI4*
AAI1*	AEO2*	AII3	AEE4
AII1	AOO2	EAO3*	AEO4*
EAE1	EAE2	EIO3	EAO4*
EAO1*	EAO2*	IAI3	EIO4
EIO1	EIO2	OAO3	IAI4

Other Kinds of Syllogism

Not all syllogisms follow the traditional rules of form. For example, we can summarize propositional deductions in only three lines. Consider the following *disjunctive syllogism*:

It is hot outside or it is cold outside.	$H \vee C$
It is not hot outside.	$\neg H$
It is cold outside.	C

A *hypothetical syllogism* resembles a *modus ponens* inference. We can represent this type of syllogism in three lines as well. For example:

If it rained this morning, then the ground is wet.	$M \rightarrow W$
It rained this morning.	M
The ground is wet.	W

QUIZ

You may refer to the text in this chapter while taking this quiz. A good score is at least 8 correct. Answers are in the back of the book.

1. **If you know that Jack is American, the rule of existential introduction allows you to conclude that**

   A. somebody is American.
   B. everyone (in the universe of discourse) is American.
   C. an equivalent statement is "One American is Jack."
   D. for any given thing, if that thing is Jack, then it is American.

2. **Which of the following formulas represents a good translation of the sentence "No mammals with colds are dogs"?**

   I. $(\forall x)\ Mx \wedge Cx \rightarrow \neg Dx$
   II. $(\forall x)\ Dx \rightarrow \neg Mx \wedge Cx$
   III. $(\forall x)\ (\exists y)\ Mx \wedge Cx \wedge Dy$
   IV. $(\exists x)\ Dx \rightarrow \neg(\forall y)\ My \wedge Cy$
   V. $\neg(\exists x)\ Mx \wedge Cx \wedge Dx$

   A. I, II, and IV
   B. Only II
   C. I and V
   D. None of the above

3. **Of the four quantifier inference rules (∃I, ∃E, ∀I, and ∀E), which involves dropping an assumption from the premises?**

   A. ∃I and ∃E (but not ∀I or ∀E)
   B. All except ∀E
   C. ∃E only
   D. None of them

4. **Suppose you observe that "Socrates was a philosopher." Your friend says, "All philosophers are bald men." These two statements could make a valid categorical syllogism proving**

   A. that all bald men are Socrates.
   B. that Socrates was a bald man.
   C. that all bald men are philosophers.
   D. nothing in particular; we can't construct a well-formed, valid syllogism with these two statements.

5. **Which of the following statements accurately describes the proof in Table 3-17?**

   A. This is a valid proof of the sequent.
   B. The proof contains an error: it does not follow the restrictions on arbitrary variables.
   C. The proof is a valid deduction, but it does not prove the sequent.
   D. The proof contains an error: line 3 does not follow from line 2.

TABLE 3-17 Proof that $(\exists x)\ Fx \wedge Gx \vdash (\forall x)\ Fx$.

Line number	Proposition	Reference	Rule	Assumption
1	$(\exists x)\ Fx \wedge Gx$		A	1
2	$Fa \wedge Ga$		A	2
3	Fa	2	$\wedge$E	2
4	Fa	1, 2, 3	$\exists$E	1
5	$(\forall x)\ Fx$	4	$\forall$I	1

6. What, if anything, can we do to correct the proof in Question 5?

A. We should reverse the sequent to obtain $(\forall x)\ Fx \vdash (\exists x)\ Fx \wedge Gx$.
B. We should change the rule at line 2 to existential elimination ($\exists$E).
C. We can't do anything to correct the proof, because the sequent is not provable.
D. We don't need to do anything, because the proof is correct as shown.

7. Consider the two-place relation of "being an ancestor," R, so that R*xy* means "*x* is an ancestor of *y*." What combination of properties accurately describes this predicate?

A. Nonsymmetric, nonreflexive, and transitive
B. Symmetric, reflexive, and transitive
C. Antisymmetric, irreflexive, and intransitive
D. Asymmetric, irreflexive, and transitive

8. Which of the following statements accurately describes the proof in Table 3-18?

A. This is a valid proof of the sequent.
B. The proof contains an error: it does not follow the restrictions on arbitrary variables.
C. The proof contains an error in the assumptions column.
D. The proof contains an error: line 5 does not follow from line 4.

TABLE 3-18 Proof that $(\forall x)\ Fx, (\exists x)\ Gx \vdash (\exists x)\ Fx \wedge Gx$.

Line number	Proposition	Reference	Rule	Assumption
1	$(\forall x)\ Fx$		A	1
2	Fa	1	$\forall$E	1
3	Ga		A	3
4	$Fa \wedge Ga$	2, 3	$\wedge$I	1, 3
5	$(\exists x)\ Fx \wedge Gx$	4	$\exists$I	1, 3
6	$(\exists x)\ Gx$		A	6
7	$(\exists x)\ Fx \wedge Gx$	3, 5, 6	$\exists$E	1, 6

9. What, if anything, can be done to correct the proof in Question 8?

A. We need to add one or more premises to the sequent and the assumptions on line 7.
B. We should reverse the rules on lines 5 and 7 so that line 5 specifies ∃E and line 7 specifies ∃I.
C. We must replace the entire assumptions column.
D. We don't need to do anything, because the proof is correct as shown.

10. Suppose someone says to you, "There is never a day that when it isn't hot outside. Therefore, it is hot outside every day." This pair of statements constitutes an example of the reasoning behind

A. one of DeMorgan's laws.
B. a hypothetical syllogism.
C. one of the laws of quantifier transformation.
D. no law, because the reasoning is fallacious.

chapter 4

A Boot Camp for Rigor

The term *rigor* means the methodical, orderly arrangement and evolution of definitions, assumptions, and resultant truths to generate a mathematical system. The material in this chapter involves basic *plane geometry*, also called *Euclidean geometry* (after *Euclid of Alexandria*, a Greek mathematician who, around 300 B.C., compiled one of the most substantial volumes ever written in mathematics and logic). Many middle schools use Euclidean geometry as a "boot camp" to give students their first taste of mathematical rigor. Some people enjoy such formality, perhaps because it offers a rational refuge from "real world chaos." Other students (alas!) simply hate it.

CHAPTER OBJECTIVES

In this chapter, you will

- Learn how to deal with terms that lack formal definitions.
- Compose basic definitions for abstract entities.
- Scrutinize the relationships between geometric objects.
- Develop a set of axioms for a logical theory.
- Prove theorems based on the axioms and definitions we have composed.

Definitions

Let's begin by stating some definitions. That's the first step in the evolution of any logical or mathematical system. We're about to launch into a rather long sequence of jargon-packed statements. Use your "logical mind" to envision their meanings.

Elementary Terms

Three objects in geometry lack formal definitions. We call them *elementary objects*. We can intuit them by comparing them with "perfect" physical objects.

- We can imagine a *point* as a ball with a radius of zero, or a brick that measures zero units along each edge. A point has a location or position that we can specify with absolute precision, but it has no volume or mass. All points are *zero-dimensional* (0D).
- We can imagine a *line* an infinitely thin, infinitely long, perfectly straight strand of wire. It extends forever in two opposite directions. A line has a position and an orientation that we can specify with absolute precision, but it has no volume or mass. All lines are *one-dimensional* (1D).
- We can think of a *plane* as an infinitely thin, perfectly flat sheet of paper that goes on forever without any edges. (I've always liked the expression "an endless frosted pane of glass without the glass.") A plane has a position and orientation that we can specify with absolute precision, but it has no volume or mass. All planes are *two-dimensional* (2D).

A line has no end points. A plane has no edges. These properties, along with the fact that a point has position but no dimension, make these three elementary objects strange indeed! You'll never hit a point with a baseball bat, slice a chunk of cheese with a line, or cross-country ski across a plane. Points, lines, and planes lack physical reality—and yet, somehow, they're *precisely* what we think they are!

Line Segment

Imagine two distinct points called P and Q, both of which lie on the same geometric line L. We define the *closed line segment PQ* as the set of all points on L between, and including, points P and Q. Figure 4-1A illustrates this definition. We draw the closed line segment's *end points* as solid dots. When you hear or read the term *line segment*, you should think of a closed line segment unless the author tells you otherwise.

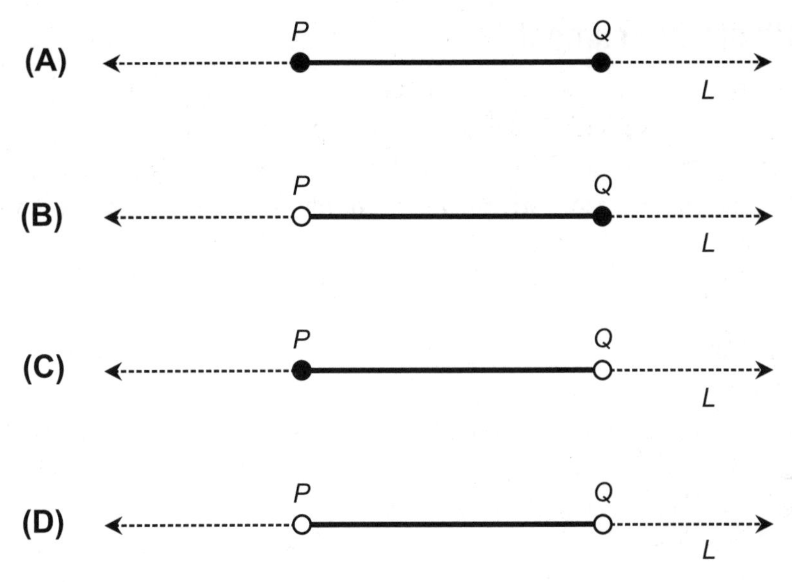

FIGURE 4-1 • At A, the closed line segment *PQ*. At B and C, half-open line segments *PQ*. At D, the open line segment *PQ*.

Half-Open Line Segment

Consider two distinct points *P* and Q, both of which lie on the same line *L*. The *half-open line segment* PQ can comprise either of two objects:

- The set of all points on *L* between *P* and Q but not including *P*, as shown in Fig. 4-1B
- The set of all points on *L* between *P* and Q but not including Q, as shown in Fig. 4-1C

In both of these cases, we draw the *excluded end points* as open circles, and the *included end points* as solid dots.

Open Line Segment

Consider two distinct points *P* and Q, both of which lie on the same line *L*. We define the *open line segment* PQ as the set of all points on *L* between, but not including, points *P* and Q, as shown in Fig. 4-1D. We draw the end points as open circles, because neither end point actually forms part of the line segment.

Length of Line Segment

Suppose that we have a closed, half-open, or open line segment *PQ* defined by distinct end points *P* and Q. We define the *length* of *PQ* as the shortest possible distance between points *P* and Q. We define the length of line segment *QP* as equal to (not the negative of!) the length of line segment *PQ*. It doesn't make any difference which way we go when we define the length of a line segment.

Closed-Ended Ray

Consider two distinct points *P* and Q, both of which lie on the same line *L*. The *closed-ended ray* PQ, also called the *closed-ended half-line* PQ, comprises the set of all points on *L* that lie on the side of point *P* that contains point Q, including point *P* itself. Figure 4-2A illustrates this scenario. When you hear or read the term *ray*, you should think of a closed-ended ray unless the author tells you otherwise.

Open-Ended Ray

Consider two distinct points *P* and Q, both of which lie on the same line *L*. The *open-ended ray* PQ, also called the *open-ended half-line* PQ, comprises the set of all points on *L* that lie on the side of point *P* that contains point Q, but not including point *P* itself, as shown in Fig. 4-2B.

Point of Intersection

Imagine a line, line segment, or ray *PQ* defined by distinct points *P* and Q. Consider a line, line segment, or ray *RS* defined by points *R* and *S*, different from points *P* and Q. Consider a point *T* that lies on both *PQ* and *RS*. We call

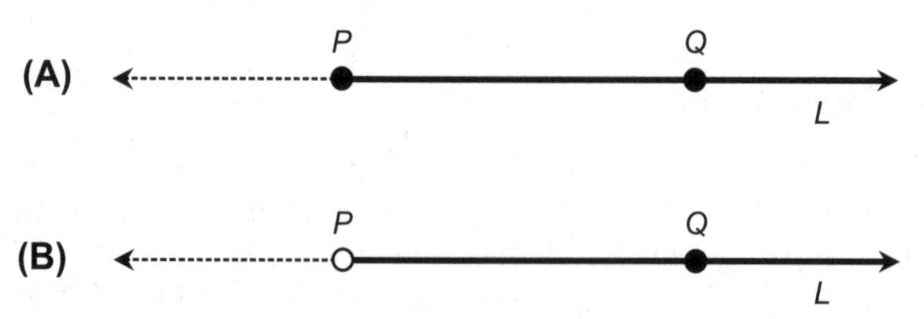

FIGURE 4-2 • At A, the closed-ended ray *PQ*. At B, the open-ended ray *PQ*.

T a *point of intersection* between *PQ* and *RS*. Alternatively, we can say that *PQ* and *RS* *intersect* at point *T*.

Collinear Points

Imagine three or more *mutually distinct* points (that means no two of the points coincide). We define the points as *collinear* if and only if they all lie on a single geometric line.

Coplanar Points

Imagine four or more mutually distinct points. We define the points as *coplanar* if and only if they all lie in a single geometric plane.

Coincident Lines

Consider four distinct points *P*, *Q*, *R*, and *S*. Consider the line *PQ*, defined by points *P* and *Q*, and the line *RS*, defined by points *R* and *S*. We define *PQ* and *RS* as *coincident lines* if and only if the four points *P*, *Q*, *R*, and *S* are collinear.

Collinear Line Segments and Rays

Consider four distinct points *P*, *Q*, *R*, and *S*. Let *PQ* represent a closed, half-open, or open line segment or ray defined by points *P* and *Q*. Let *RS* represent a closed, half-open, or open line segment or ray defined by points *R* and *S*. We define *PQ* and *RS* as *collinear* if and only if points *P*, *Q*, *R*, and *S* are collinear.

Still Struggling

Are you wondering why the word "let" appears in the previous definition? Does the statement "Let *PQ* represent ..." seem strange? You'll often encounter this sort of language in mathematical papers, articles, and presentations. When an author tells you to "let" things be this way or that way, you should imagine, suppose, or consider that things are this way or that, just for the sake of the argument at hand. You should "let things be so," setting the scene in your mind for statements, problems, definitions, or proofs to follow.

Transversal

Consider four distinct points *P*, *Q*, *R*, and *S*. Let *PQ* be the line defined by points *P* and *Q*. Let *RS* be the line defined by points *R* and *S*. Suppose that line *PQ* and line *RS* are not coincident. Let *L* be a line that intersects both line *PQ* and line *RS*. We call line *L* a *transversal* of lines *PQ* and *RS*.

Parallel Lines

Consider four distinct points *P*, *Q*, *R*, and *S*. Let *PQ* be the line defined by points *P* and *Q*. Let *RS* be the line defined by points *R* and *S*. Suppose that lines *PQ* and *RS* lie in the same plane, but are not coincident. We call *PQ* and *RS* *parallel lines* if and only if there exists no point at which they intersect.

Parallel Line Segments

Consider four distinct points *P*, *Q*, *R*, and *S*. Let line *PQ* be the line defined by points *P* and *Q*. Let line segment *PQ* be a closed, half-open, or open line segment contained in line *PQ*, with end points *P* and *Q*. Let *RS* be the line defined by points *R* and *S*. Suppose that lines *PQ* and *RS* lie in the same plane, but are not coincident. Let line segment *RS* be a closed, half-open, or open line segment contained in line *RS*, with end points *R* and *S*. We say that *PQ* and *RS* constitute *parallel line segments* if and only if lines *PQ* and *RS* are parallel lines.

Figure 4-3A shows an example of parallel line segments, one closed and the other half-open. Drawing B shows an example of two half-open line segments that aren't parallel. Drawing C shows an example of two open line segments that aren't parallel. In the situation of Fig. 4-3A, there exists no point of intersection common to both lines. In the situation of Fig. 4-3B, a point *T* exists that's common to both lines, although *T* does not lie on either line segment. In the situation of Fig. 4-3C, there exists a point *T* common to both lines, and *T* lies on both line segments.

Parallel Rays

Consider four distinct points *P*, *Q*, *R*, and *S*. Let *PQ* be the line defined by points *P* and *Q*. Let ray *PQ* be a closed-ended or open-ended ray contained in line *PQ*, with end point *P*. Let *RS* be the line defined by points *R* and *S*. Suppose that lines *PQ* and *RS* lie in the same plane, but are not coincident. Let ray *RS* be a closed-ended or open-ended ray contained in line *RS*, with end

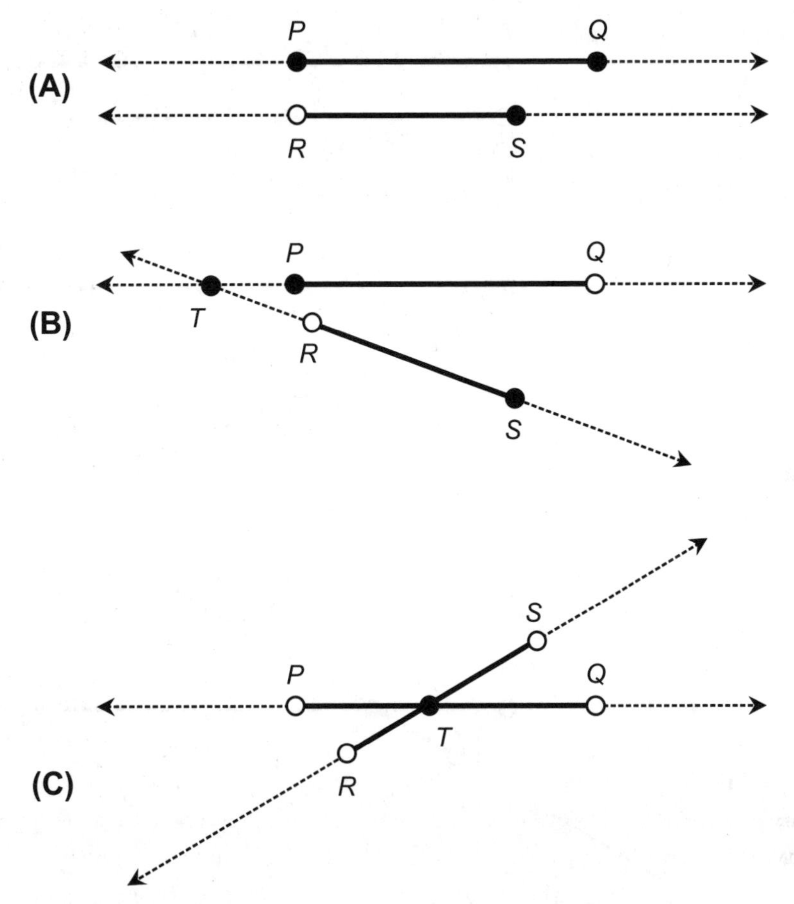

FIGURE 4-3 • At A, line segments *PQ* and *RS* are parallel, because lines *PQ* and *RS* lie in the same plane and do not intersect. At B and C, line segments *PQ* and *RS* are not parallel, because lines *PQ* and *RS* intersect at a point *T*.

point *R*. We say that rays *PQ* and *RS* constitute *parallel rays* if and only if lines *PQ* and *RS* are parallel lines.

In Fig. 4-4, drawing A shows an example of parallel rays, one closed-ended and the other open-ended. Drawing B shows an example of two rays, one closed-ended and the other open-ended, that aren't parallel. Drawing C shows an example of two rays, both open-ended, that aren't parallel. In the situation of Fig. 4-4A, there exists no point of intersection common to both lines. In the situation of Fig. 4-4B, there exists a point *T* common to both lines, although *T* does not lie on either ray. In the situation of Fig. 4-4C, there exists a point *T* common to both lines, and *T* lies on both rays.

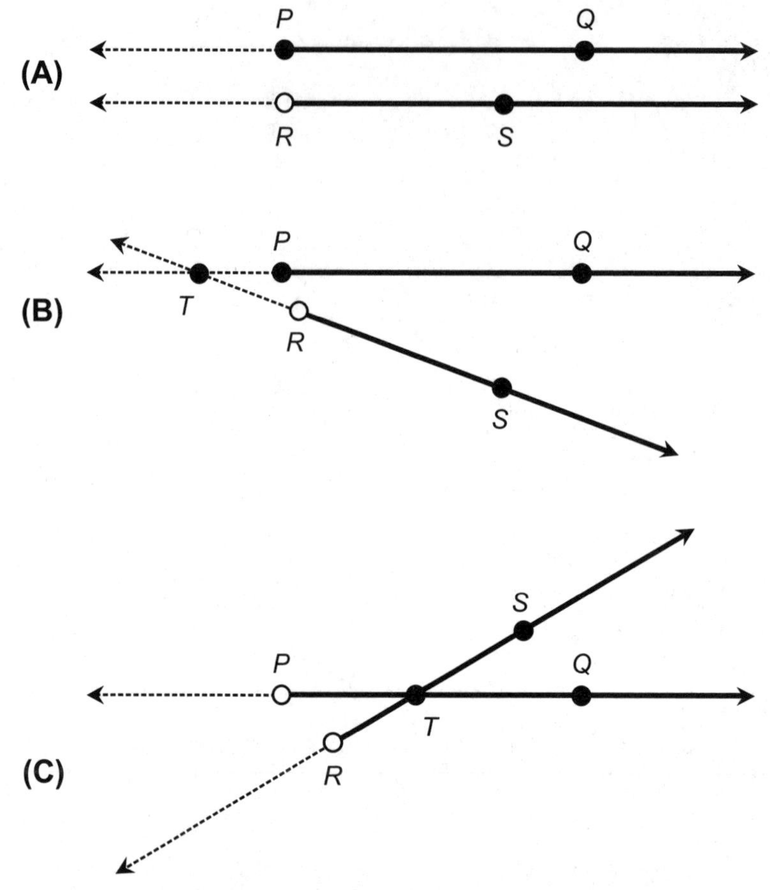

FIGURE 4-4 • At A, rays *PQ* and *RS* are parallel, because lines *PQ* and *RS* lie in the same plane and do not intersect. At B and C, rays *PQ* and *RS* are not parallel, because lines *PQ* and *RS* intersect at a point *T*.

Angle

Consider three distinct points *P*, Q, and *R*. Suppose that Q*P* and Q*R* are rays or line segments, both of which have the same end point Q. We say that the two rays or line segments and their common end point constitute an *angle* denoted ∠*P*Q*R*. We call the rays or line segments Q*P* and Q*R* the *sides* of ∠*P*Q*R*, and we call point Q the *vertex* of ∠*P*Q*R*. Figure 4-5 illustrates an example. Unless otherwise specified, we rotate counterclockwise when we define an angle.

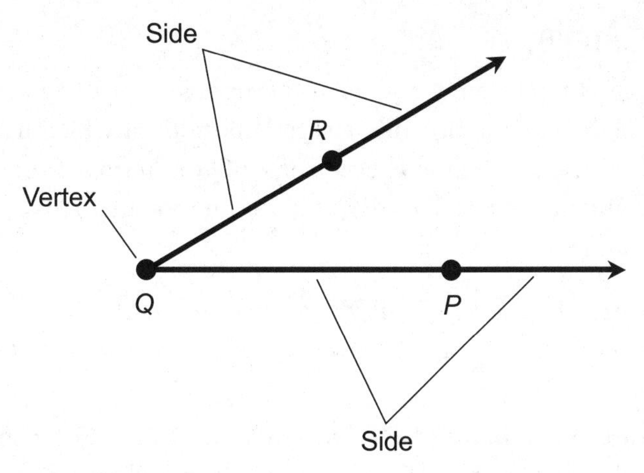

FIGURE 4-5 • An angle comprises two rays or line segments that share an end point.

PROBLEM 4-1

What's ambiguous about Fig. 4-5?

SOLUTION

The two rays shown in Fig. 4-5 can actually define two different angles. The angle that most people imagine when they look at the drawing goes counterclockwise "the short way around" from ray *QP* to ray *QR*. The other angle, which most people don't think of at first, goes counterclockwise "the long way around" from ray *QR* to ray *QP*. To distinguish this angle from $\angle PQR$, we call it $\angle RQP$.

PROBLEM 4-2

What happens if we think of angles as rotating in the clockwise sense, rather than counterclockwise? Do they constitute legitimate angles, too? If so, how can we describe them in the situation of Fig. 4-5?

SOLUTION

We can consider angles as rotating in the clockwise sense, defining such an angle as the *negative* of the angle having the same sides going counterclockwise. In the scenario of Fig. 4-5 rotating clockwise, $\angle RQP$ goes the short way around, and $\angle PQR$ goes the long way around.

Measure of Angle

Consider three distinct points P, Q, and R. Suppose that QP and QR are distinct, closed-ended rays or closed line segments, both of which have the same end point Q. We can define the *measure* of $\angle PQR$ in terms of the portion of a complete revolution described by $\angle PQR$. Two common methods exist for doing this task, as follows:

- The *measure in degrees* of $\angle PQR$, symbolized $m^\circ\angle PQR$, constitutes the fractional part of a complete revolution described by $\angle PQR$, multiplied by 360.
- The *measure in radians* of $\angle PQR$, symbolized $m\angle PQR$, constitutes the fractional part of a complete revolution described by $\angle PQR$, multiplied by 2π, where π represents the ratio of a circle's circumference to its diameter (an irrational number equal to approximately 3.14159).

When we want to denote the measure of an angle as a variable, we can use a lowercase italic English letter such as x or y, or a lowercase italic Greek letter such θ or ϕ.

Straight Angle

Consider three distinct points P, Q, and R. Let QP and QR be rays or line segments, both of which have the same end point Q, and which define $\angle PQR$. We call $\angle PQR$ a *straight angle* if and only if points P, Q, and R are collinear and Q lies between P and R.

Straight Angle (first alternate definition)

An angle constitutes a *straight angle* if and only if it represents precisely half of a complete revolution.

Straight Angle (second alternate definition)

An angle constitutes a *straight angle* if and only its measure equals precisely 180°.

Straight Angle (third alternate definition)

An angle constitutes a *straight angle* if and only its measure equals precisely π radians (rad).

Supplementary Angles

We call two angles *supplementary angles* if and only if the sum of their measures equals the measure of a straight angle.

Right Angle

Consider five distinct points P, Q, R, S, and T, all of which lie in the same plane, but not all of which are collinear. Consider ray TP, ray TQ, ray TR, and ray TS. Suppose that both of the following statements hold true:

- The rays define $\angle PTQ$, $\angle QTR$, $\angle RTS$, and $\angle STP$
- All four angles $\angle PTQ$, $\angle QTR$, $\angle RTS$, and $\angle STP$ have equal measure

In this situation, we define each of the four angles $\angle PTQ$, $\angle QTR$, $\angle RTS$, and $\angle STP$ as a *right angle*.

Right Angle (first alternate definition)

An angle constitutes a *right angle* if and only if its measure equals exactly half the measure of a straight angle.

Right Angle (second alternate definition)

An angle constitutes a *right angle* if and only its measure equals precisely 90°.

Straight Angle (third alternate definition)

An angle constitutes a *right angle* if and only its measure equals precisely $\pi/2$ rad.

Complementary Angles

We define two angles as *complementary angles* if and only if the sum of their measures equals the measure of a right angle.

Perpendicular Lines, Line Segments, and Rays

Consider four distinct points P, Q, R, and S. We say that a line, line segment, or ray PQ is *perpendicular* to a line, line segment, or ray RS if and only if both of the following statements hold true:

- Lines PQ and RS intersect at one and only one point T, and
- One of the angles at point T, formed by the intersection of line PQ and line RS, constitutes a right angle.

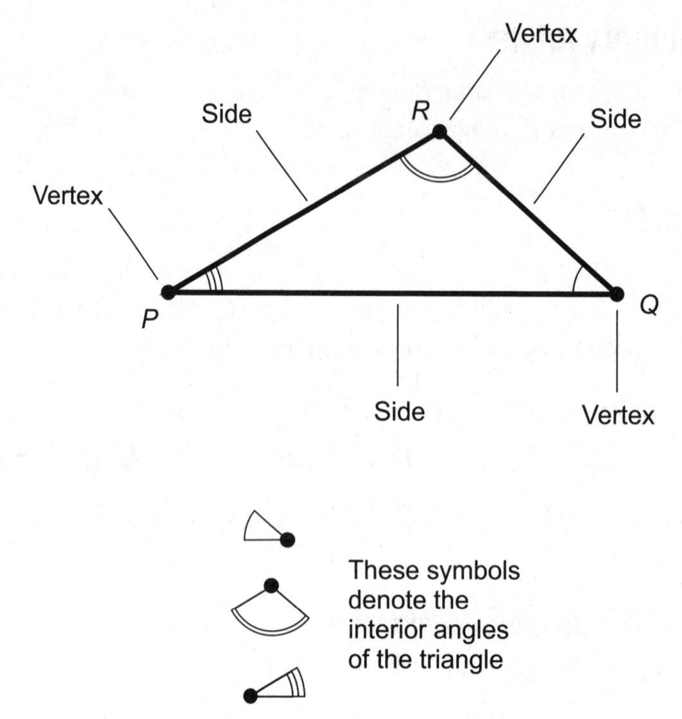

FIGURE 4-6 • A triangle comprises three points and the line segments that connect them.

Triangle

Consider three distinct points P, Q, and R. Imagine the closed line segment PQ, the closed line segment QR, and the closed line segment RP. These three line segments, along with the points P, Q, and R, constitute a *triangle* denoted as ΔPQR. Figure 4-6 illustrates an example in which line segment PQ, line segment QR, and line segment RP form the *sides* of ΔPQR. We call the points P, Q, and R the *vertices* of ΔPQR. We call the angles $\angle RQP$, $\angle PRQ$, and $\angle QPR$ the *interior angles* of ΔPQR.

Geometers often mark the interior angles of a triangle by drawing sets of concentric arcs centered on the vertices and connecting pairs of adjacent sides. You might wish to place an arrow at either end of such an arc to specify the rotational sense of the angle. If you see arcs without arrows denoting angles, or if you draw the arcs without arrows, then you should imagine the angle rotation as going counterclockwise.

Included Angle

Imagine a triangle's interior angle whose vertex lies at the point where two specific sides intersect. We call such an angle an *included angle*. For example, in Fig. 4-6, $\angle QPR$ constitutes the included angle between the sides PQ and PR.

Included Side

In a triangle, an *included side* is a side whose end points constitute the vertices of two specific angles. For example, in Fig. 4-6, line segment PQ forms the included side between $\angle RQP$ and $\angle QPR$.

Isosceles Triangle

Imagine a triangle defined by three distinct points P, Q, and R. Let p be the length of the side opposite point P. Let q be the length of the side opposite point Q. Let r be the length of the side opposite point r. Suppose that *at least one* of the following equations holds true:

$$p = q$$
$$q = r$$
$$p = r$$

We call a triangle of this type an *isosceles triangle*. It has at least one pair of sides whose lengths are equal.

Equilateral Triangle

Imagine a triangle defined by three distinct points P, Q, and R. Let p be the length of the side opposite point P. Let q be the length of the side opposite point Q. Let r be the length of the side opposite point R. Suppose that *all three* of the following equations hold true:

$$p = q$$
$$q = r$$
$$p = r$$

We can state this condition in more concise terms as the single equation

$$p = q = r$$

We call a triangle of this type an *equilateral triangle*. All three sides have the same length.

Right Triangle

We call a triangle a *right triangle* if and only if one of its interior angles is a right angle.

Similar and Congruent Triangles

Let's look at some definitions that apply especially to triangles. In our theorem-proving exercises (we're getting there!), we'll encounter the notions of *similarity* and *congruence*, and in particular, a property called *direct congruence*.

Direct Similarity

Consider three distinct points P, Q, and R as shown in Fig. 4-7A. Let ΔPQR be a triangle defined by proceeding counterclockwise from point P to point Q, from point Q to point R, and from point R to point P. Let p be the length of the side opposite point P. Let q be the length of the side opposite point Q.

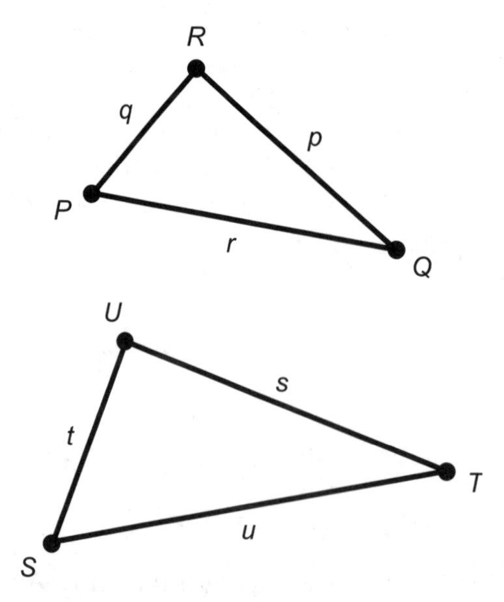

FIGURE 4-7A • Directly similar triangles.

Let r be the length of the side opposite point R. Now consider three more distinct points S, T, and U. Suppose that all six points P, Q, R, S, T, and U are coplanar. Let ΔSTU be a triangle distinct from ΔPQR, proceeding counterclockwise from point S to point T, from point T to point U, and from point U to point S. Let s be the length of the side opposite point S. Let t be the length of the side opposite point T. Let u be the length of the side opposite point U. We say that ΔPQR and ΔSTU are *directly similar* if and only if the lengths of their corresponding sides, as we proceed in the same direction around either triangle, exist in a constant ratio; that is, if and only if the following equation holds true:

$$p/s = q/t = r/u$$

In addition, ΔPQR and ΔSTU are directly similar if and only all three of the following equations hold true:

$$\text{m}\angle QPR = \text{m}\angle TSU$$
$$\text{m}\angle PRQ = \text{m}\angle SUT$$
$$\text{m}\angle RQP = \text{m}\angle UTS$$

We can informally describe direct similarity as follows: If we enlarge or reduce one triangle to exactly the correct extent, and then rotate that triangle clockwise or counterclockwise to exactly the correct extent (without flipping it over!), we can place that triangle directly on top of the other one, so that the two triangles coincide.

The direct similarity symbol looks like a wavy minus sign, known as a *tilde* (~). If we have two directly similar triangles ΔPQR and ΔSTU, we can symbolize their similarity as follows:

$$\Delta PQR \sim \Delta STU$$

Inverse Similarity

Consider three distinct points P, Q, and R as shown in Fig. 4-7B. Let ΔPQR be a triangle defined by proceeding counterclockwise from point P to point Q, from point Q to point R, and from point R to point P. Let p be the length of the side opposite point P. Let q be the length of the side opposite point Q.

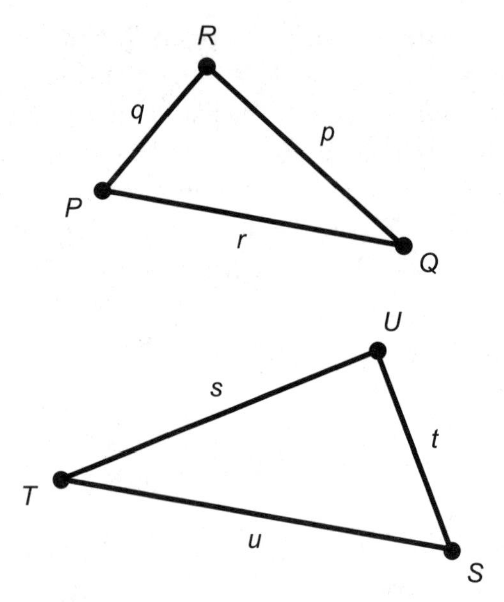

FIGURE 4-7B • Inversely similar triangles.

Let r be the length of the side opposite point R. Let S, T, and U be distinct points. Suppose that points P, Q, R, S, T, and U are all coplanar. Let ΔSTU be a triangle distinct from ΔPQR, proceeding *clockwise* (not counter-clockwise!) from point S to point T, from point T to point U, and from point U to point S. Let s be the length of the side opposite point S. Let t be the length of the side opposite point T. Let u be the length of the side opposite point U. We define ΔPQR and ΔSTU as *inversely similar* if and only if the lengths of their corresponding sides, as we proceed in *opposite directions* around the triangles, exist in a constant ratio; that is, if and only if the following equation holds true:

$$p/s = q/t = r/u$$

In addition, ΔPQR and ΔSTU are inversely similar if and only if all three of the following equations hold true:

$$\text{m}\angle QPR = \text{m}\angle UST$$

$$\text{m}\angle PRQ = \text{m}\angle TUS$$

$$\text{m}\angle RQP = \text{m}\angle STU$$

An alternative, but informal, way to describe inverse similarity is to imagine that, if one triangle is flipped over, enlarged or reduced by just the right amount, and finally rotated clockwise or counterclockwise to just the right extent, we can "paste" one triangle on top of the other so that they coincide.

The inverse similarity symbol is not universally agreed-upon. Let's use a tilde followed by a minus sign (~–). If distinct triangles ΔPQR and ΔSTU are inversely similar, then we can symbolize the fact as follows:

$$\Delta PQR \sim\!- \Delta STU$$

TIP ***When you read or hear that two triangles are "similar," you'll want to know for certain what the author means! Sometimes the term*** **similarity** ***refers only to direct similarity, but in some texts it can refer to either direct similarity or inverse similarity. We can avoid confusion by using the full terminology all the time.***

Direct Congruence

Consider three distinct points P, Q, and R as shown in Fig. 4-8A. Let ΔPQR be a triangle defined by proceeding counterclockwise from point P to point Q, from point Q to point R, and from point R to point P. Let p be the length of the side opposite point P. Let q be the length of the side opposite point Q. Let r be the length of the side opposite point R. Now consider three more distinct points S, T, and U. Suppose that points P, Q, R, S, T, and U are all coplanar. Let ΔSTU be a triangle distinct from ΔPQR, proceeding counterclockwise from point S to point T, from point T to point U, and from point U to point S. Let s be the length of the side opposite point S. Let t be the length of the side opposite point T. Let u be the length of the side opposite point U. Then ΔPQR and ΔSTU are *directly congruent* if and only if the lengths of their corresponding sides, as we proceed in the same direction around either triangle, are equal; that is, if and only if all three of the following equations hold true:

$$p = s$$

$$q = t$$

$$r = u$$

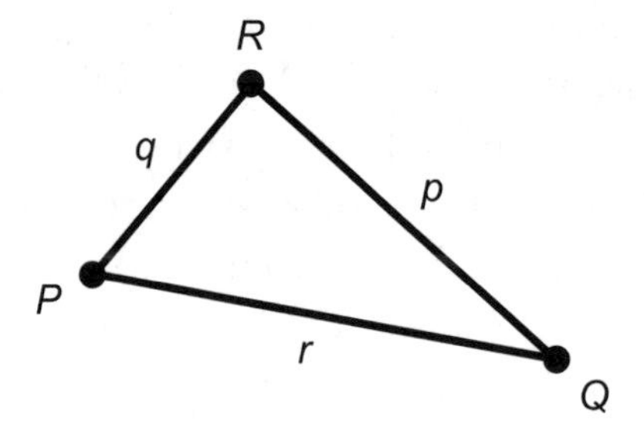

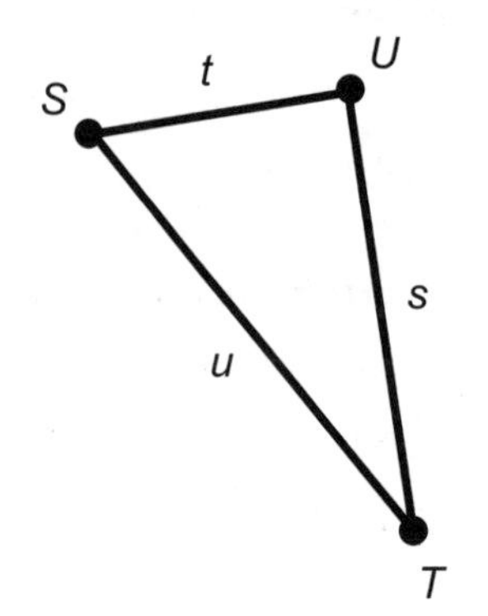

FIGURE 4-8A • Directly congruent triangles.

In addition, if ΔPQR and ΔSTU are directly congruent then all three of the following equations hold true:

$$\text{m}\angle QPR = \text{m}\angle TSU$$

$$\text{m}\angle PRQ = \text{m}\angle SUT$$

$$\text{m}\angle RQP = \text{m}\angle UTS$$

We can informally describe direct congruence as follows: If we rotate one triangle clockwise or counterclockwise to exactly the correct extent (without flipping it over!), we can place it directly on top of the other triangle, so that the two triangles coincide.

The direct congruence symbol looks like a triple-barred equals sign (≡). If distinct triangles ΔPQR and ΔSTU are directly congruent, we symbolize the fact by writing

$$\Delta PQR \equiv \Delta STU$$

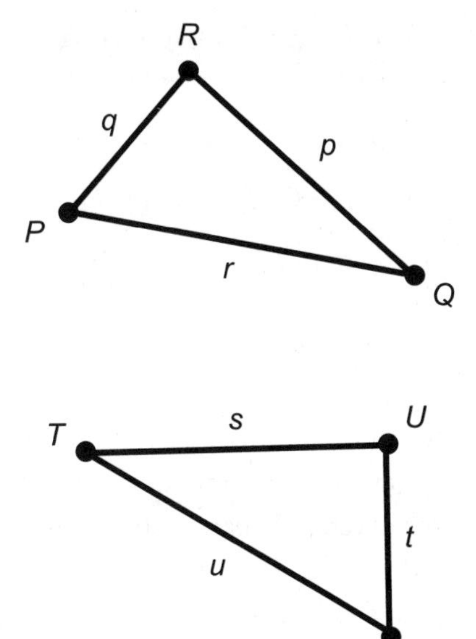

FIGURE 4-8B • Inversely congruent triangles.

Inverse Congruence

Consider three distinct points P, Q, and R as shown in Fig. 4-8B. Let ΔPQR be a triangle defined by proceeding counterclockwise from point P to point Q, from point Q to point R, and from point R to point P. Let p be the length of the side opposite point P. Let q be the length of the side opposite point Q. Let r be the length of the side opposite point R. Consider three more distinct points S, T, and U. Suppose that points P, Q, R, S, T, and U are all coplanar. Let ΔSTU be a triangle distinct from ΔPQR, proceeding *clockwise* (not counterclockwise!) from point S to point T, from point T to point U, and from point U to point S. Let s be the length of the side opposite point S. Let t be the length of the side opposite point T. Let u be the length of the side opposite point U. Then ΔPQR and ΔSTU are *inversely congruent* if and only if the lengths of their corresponding sides, as we proceed in opposite directions around the triangles, are equal; that is, if and only if all three of the following equations hold true:

$$p = s$$
$$q = t$$
$$r = u$$

In addition, if ΔPQR and ΔSTU are inversely congruent then all three of the following equations hold true:

$$m\angle QPR = m\angle UST$$

$$m\angle PRQ = m\angle TUS$$

$$m\angle RQP = m\angle STU$$

An alternative, but informal, way to describe inverse congruence is to imagine that, if we flip one triangle over and then rotate it clockwise or counterclockwise to just the right extent, we can paste the two triangles on top of each other so that they coincide.

The inverse congruence symbol is not universally agreed-upon. Let's use a triple-barred equals sign followed by a minus sign ($\equiv-$). If distinct triangles ΔPQR and ΔSTU are inversely congruent, then, we can symbolize the fact by writing

$$\Delta PQR \equiv- \ \Delta STU$$

TIP ***When you read or hear that two triangles are* congruent triangles, *you had better make certain that you know what the author means! Sometimes the term* congruence *refers only to direct* congruence, *but in some texts it can refer to either direct congruence or inverse congruence. As with similarity, let's always use the full terminology!***

Two Crucial Facts

Here are two important things you should remember about directly congruent triangles. These facts will help reinforce, in your mind, the logical meaning of the term.

- If two triangles are directly congruent, then their corresponding sides have equal lengths as you proceed around both triangles in the same direction. The converse also holds true. If two triangles have corresponding sides with equal lengths as you proceed around them both in the same direction, then the two triangles are directly congruent.

- If two triangles are directly congruent, then their corresponding interior angles have equal measures as you proceed around both triangles in the same direction. The converse, however, does not necessarily hold true. Two triangles can have corresponding interior angles with equal measures when you proceed around them both in the same direction, and yet the two triangles are not directly congruent.

Two More Crucial Facts

Here are two "mirror images" of the facts just stated. They concern inversely congruent triangles. Note the subtle differences in wording!

- If two triangles are inversely congruent, then their corresponding sides have equal lengths as you proceed around the triangles in opposite directions. The converse also holds true. If two triangles have corresponding sides with equal lengths as you proceed around them in opposite directions, then the two triangles are inversely congruent.
- If two triangles are inversely congruent, then their corresponding interior angles have equal measures as you proceed around the triangles in opposite directions. The converse, however, does not necessarily hold true. Two triangles can have corresponding interior angles with equal measures as you proceed around them in opposite directions, and yet the two triangles are not inversely congruent.

PROBLEM 4-3

Explain in terms of side lengths why any two equilateral triangles are directly similar.

SOLUTION

Any two equilateral triangles are directly similar because the ratio of their corresponding sides, proceeding counterclockwise around both triangles, is a constant. Suppose ΔPQR is an equilateral triangle with sides of length m. Suppose that ΔSTU is an equilateral triangle with sides of length n. Then the ratio of the lengths of their corresponding sides, proceeding counterclockwise around ΔPQR from point P and counterclockwise around ΔSTU from point S, is always m/n. The triangles are therefore directly similar by definition.

PROBLEM 4-4

Explain in terms of side lengths why any two equilateral triangles are inversely similar.

SOLUTION

Any two equilateral triangles are inversely similar because the ratio of their corresponding sides, proceeding counterclockwise around one of them and clockwise around the other, is a constant. Suppose that ΔPQR is an equilateral triangle with sides of length *m*; also suppose that ΔSTU is an equilateral triangle with sides of length *n*. Then the ratio of the lengths of their corresponding sides, proceeding counterclockwise around ΔPQR from point *P* and clockwise around ΔSTU from point *S*, always equals m/n. (In fact, the ratio of the length of *any* side of ΔPQR to the length of *any* side of ΔSTU equals m/n.) The triangles are therefore inversely similar by definition.

Axioms

We've armed ourselves with a substantial supply of definitions! However, we still need some axioms if we want to prove anything. Several axioms follow. Do you remember any of them from your basic geometry course? Some of these axioms come straight from Euclid.

The Two-Point Axiom

Any two distinct points *P* and *Q* can be connected by a straight line segment.

The Extension Axiom

Any straight line segment, defined by distinct points *P* and *Q*, can be extended indefinitely and continuously to form a straight line defined by points *P* and *Q*.

The Right Angle Axiom

All right angles have the same measure.

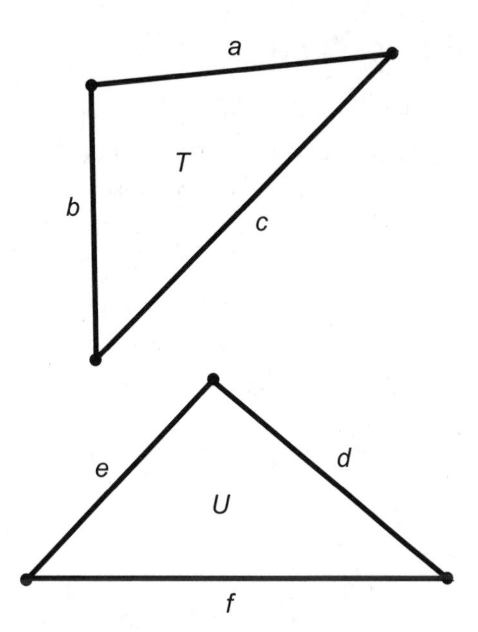

FIGURE 4-9 • The side-side-side (SSS) axiom. Triangles *T* and *U* are directly congruent if and only if $a = d$, $b = e$, and $c = f$.

The Parallel Axiom

Given a straight line *L* and a point *P* that does not lie on *L*, there exists one and only one straight line *M* that passes through *P* and runs parallel to *L*.

The Side-Side-Side (SSS) Axiom

Two triangles *T* and *U* are directly congruent *if and only if* their corresponding sides have identical lengths as you proceed around them in the same direction. (Figure 4-9 can help you visualize this principle, where $a = d$, $b = e$, and $c = f$.)

As an extension of this axiom, two triangles *T* and *U* are inversely congruent *if and only if* their corresponding sides have identical lengths as you proceed around them in the opposite directions.

TIP ***Do you think that the SSS axiom replicates the definitions that we've already formulated for direct and inverse congruence? If so, you're right! We state this axiom here because, in some texts, the definitions of direct and inverse congruence are less precise, stating only the general notions about size and shape.***

The Side-Angle-Side (SAS) Axiom

Suppose that two triangles T and U have pairs of corresponding sides of identical lengths as you proceed around the triangles in the same direction. Also, suppose that the included angles between those corresponding sides have identical measures. In that case, triangles T and U are directly congruent. Conversely, if T and U are directly congruent triangles, then T and U have pairs of corresponding sides of identical lengths as you proceed around the triangles in the same direction, and the included angles between those corresponding sides have identical measures. (Figure 4-10 can help you visualize this principle, where $a = c$, $x = y$, and $b = d$.)

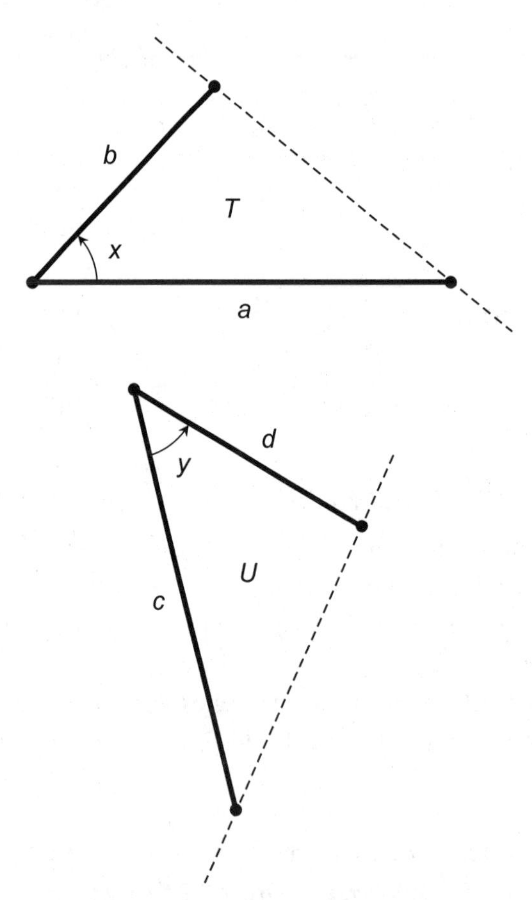

FIGURE 4-10 • The side-angle-side (SAS) axiom. Triangles T and U are directly congruent if and only if $a = c$, $x = y$, and $b = d$.

As an extension of this axiom, suppose that two triangles *T* and *U* have pairs of corresponding sides of identical lengths as you proceed around the triangles in opposite directions. Also, suppose that the included angles between those corresponding sides have identical measures. In that case, triangles *T* and *U* are inversely congruent. Conversely, if *T* and *U* are inversely congruent triangles, then *T* and *U* have pairs of corresponding sides of identical lengths as you proceed around the triangles in opposite directions, and the included angles between those corresponding sides have identical measures.

The Angle-Side-Angle (ASA) Axiom

Suppose that two triangles *T* and *U* have pairs of corresponding angles of identical measures as you proceed around the triangles in the same direction. Also, suppose that the included sides between those corresponding angles have identical lengths. In that case, triangles *T* and *U* are directly congruent. Conversely, if *T* and *U* are directly congruent triangles, then *T* and *U* have pairs of corresponding angles of identical measures as you proceed around the triangles in the same direction, and the included sides between those corresponding angles have identical lengths. (Figure 4-11 can help you visualize this. In this example, $w = y$, $a = b$, and $x = z$.)

As an extension to this axiom, suppose that two triangles *T* and *U* have pairs of corresponding angles of identical measures as you proceed around the triangles in opposite directions. Also, suppose the included sides between those corresponding angles have identical lengths. Then *T* and *U* are inversely congruent. Conversely, if *T* and *U* are inversely congruent triangles, then *T* and *U* have pairs of corresponding angles of identical measures as you proceed around the triangles in opposite directions, and the included sides between those corresponding angles have identical lengths.

The Side-Angle-Angle (SAA) Axiom

Let *T* and *U* be triangles. Suppose *T* and *U* have pairs of corresponding angles of identical measures as you proceed around the triangles in the same direction. Suppose the corresponding sides, one of whose end points constitutes the vertex of the first-encountered angle in either triangle, have identical lengths. Then *T* and *U* are directly congruent. Conversely, if *T* and *U* are directly congruent triangles, then *T* and *U* have pairs of corresponding angles of identical measures as you proceed around the triangles in the same direction, and the

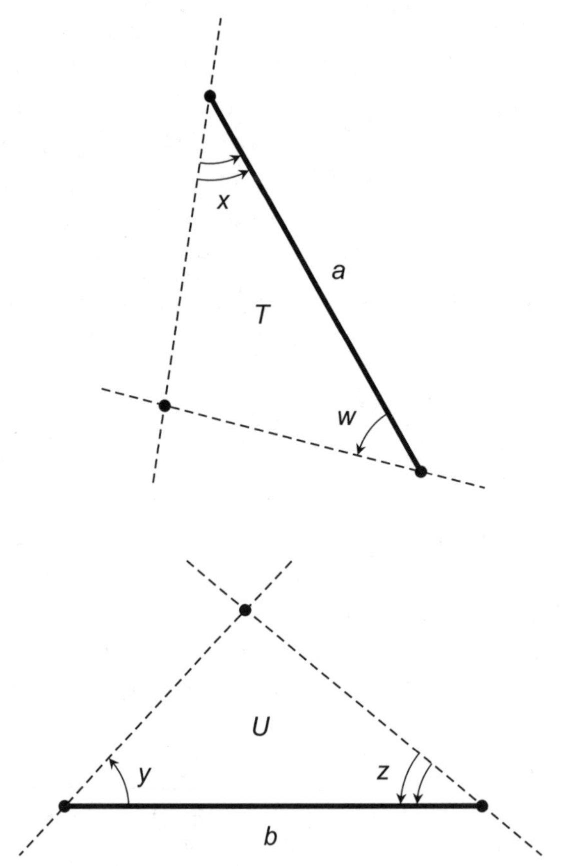

FIGURE 4-11 • The angle-side-angle (ASA) axiom. Triangles *T* and *U* are directly congruent if and only if $w = y$, $a = b$, and $x = z$.

corresponding sides whose end points constitute the vertices of the first-encountered angles have identical lengths. (Figure 4-12 can help you visualize this principle, where $a = b$, $w = y$, and $x = z$.)

As an extension to this axiom, suppose that two triangles *T* and *U* have pairs of corresponding angles of identical measures as you proceed around the triangles in opposite directions. Suppose that the corresponding sides, one of whose end points constitutes the vertex of the first-encountered angle in either triangle, have identical lengths. In that case, *T* and *U* are inversely congruent. Conversely, if *T* and *U* are inversely congruent triangles, then *T* and *U* have pairs of corresponding angles of identical measures as you proceed around the triangles in opposite directions, and the corresponding sides whose end points constitute the vertices of the first-encountered angles have identical lengths.

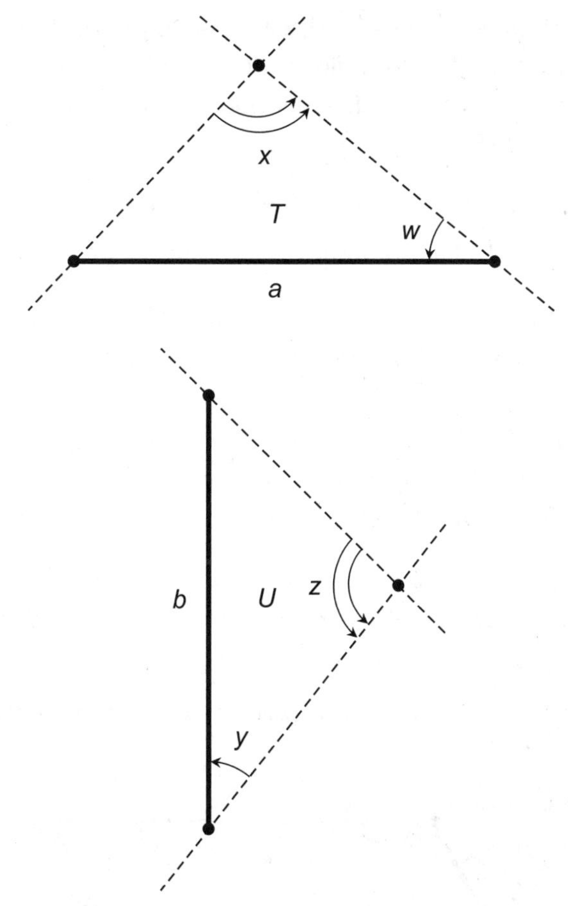

FIGURE 4-12 • The side-angle-angle (SAA) axiom. Triangles T and U are directly congruent if and only if $a = b$, $w = y$, and $x = z$.

Some Proofs at Last

We're finally ready to prove a few theorems! In every case, we'll state the proof in verbal form, and also portray it as a *statement/reason* (S/R) *table*. Let's state our theorems as "problems," and work out the proofs as "solutions."

PROBLEM 4-5

Consider four mutually distinct points *P, Q, R,* and *S*, all of which lie in the same plane. Suppose that the lines *PQ, QR, RS,* and *SP* are all mutually distinct (i.e., no two of them coincide). Consider line segments *PQ, QR, RS,*

and *SP*, each of which lies on the line having the same name. The line segments form a four-sided figure *PQRS*, with vertices at points *P*, *Q*, *R*, and *S*, in that order proceeding counterclockwise. Suppose that both of the following statements hold true:

- Line segment *PQ* has the same length as line segment *SR*
- Line segment *SP* has the same length as line segment *RQ*

Now imagine a line segment *SQ*, which divides the figure *PQRS* into two triangles, ΔSPQ and ΔQRS. Prove that $\Delta SPQ \equiv \Delta QRS$.

Figure 4-13 provides a generic illustration of this situation. Let's assign corresponding sides for the triangles, proceeding counterclockwise around the triangle in either case:

- Line segment *SP* in ΔSPQ corresponds to line segment *QR* in ΔQRS
- Line segment *PQ* in ΔSPQ corresponds to line segment *RS* in ΔQRS
- Line segment *QS* in ΔSPQ corresponds to line segment *SQ* in ΔQRS

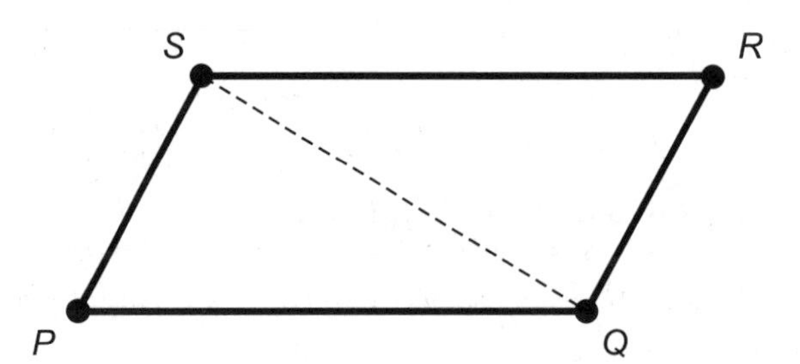

FIGURE 4-13 • A line segment that divides a polygon into two triangles. Illustration for Problem 4-5 and its solution.

We've been told that line segment *PQ* has the same length as line segment *SR*. The length of a line segment does not depend on the direction we go. Therefore, line segment *PQ* in ΔSPQ has the same length as line segment *RS* in ΔQRS. These line segments form corresponding sides in the triangles.

We've also been told that line segment *SP* has the same length as line segment *RQ*. Therefore, line segment *SP* in ΔSPQ has the same length as line segment *QR* in ΔQRS. These line segments form corresponding sides in the triangles.

TABLE 4-1 An S/R version of the proof demonstrated in the solution of Problem 4-5 and Fig. 4-13.

Statement	Reason
Line segment *SP* in ΔSPQ corresponds to line segment *QR* in ΔQRS.	We assign them that way.
Line segment *PQ* in ΔSPQ corresponds to line segment *RS* in ΔQRS.	We assign them that way.
Line segment *QS* in ΔSPQ corresponds to line segment *SQ* in ΔQRS.	We assign them that way.
Line segment *PQ* has the same length as line segment *SR*.	Given.
Line segment *SP* has the same length as line segment *RQ*.	Given.
Line segment *QS* has the same length as line segment *SQ*.	This comes from the definition of the length of a line segment: it is the same in either direction.
Corresponding sides of ΔSPQ and ΔQRS have identical lengths expressed counterclockwise.	This is based on the above statements, and on the way we have assigned corresponding sides.
$\Delta SPQ \equiv \Delta QRS$.	SSS axiom.

Finally, from the definition of the length of a line segment, we know that line segment *QS* in ΔSPQ has the same length as line segment *SQ* in ΔQRS. These line segments constitute corresponding sides (which also happen to coincide) in the triangles.

We've now shown that the corresponding sides of ΔSPQ and ΔQRS have identical lengths when we proceed around the triangles in the same direction (counterclockwise in this case). Therefore, according to the SSS axiom, we can conclude that that $\Delta SPQ \equiv \Delta QRS$. Table 4-1 shows the S/R version of this proof.

PROBLEM 4-6

Consider four distinct points *P*, *Q*, *R*, and *S*, all of which lie in the same plane. Suppose that the four lines *PQ*, *RQ*, *SR*, and *SP* are all mutually distinct. Consider the four line segments *PQ*, *RQ*, *SR*, and *SP*, each of which lies on the line having the same name. Imagine that this set of line segments forms

a four-sided figure *PQRS*, with vertices at points *P*, *Q*, *R*, and *S*, in that order proceeding counterclockwise. Suppose that the following statements both hold true:

- Line segment *PQ* has the same length as line segment *SR*
- Line segment *RQ* has the same length as line segment *SP*

Consider lines *PQ* and *SR*, the extensions of line segments *PQ* and *SR*, respectively. Also consider line *SQ*, the extension of line segment *SQ*. Prove that $m\angle QSR = m\angle SQP$.

SOLUTION

Figure 4-14 provides us with a generic diagram of this situation. Because we've solved Problem 4-5, we can use its result as a theorem. In the situation at hand, we know that $\Delta SPQ \equiv \Delta QRS$. When two triangles are directly congruent, then the counterclockwise measures of their corresponding angles, as we proceed in the same direction around either triangle, are equal. From the description of this situation (and Fig. 4-14), we can see that $\angle QSR$ and $\angle SQP$ constitute corresponding angles; the first angle lies inside ΔQRS, and the second angle lies inside ΔSPQ. We express the angle measures in the counterclockwise sense, so they are equal; that is, $m\angle QSR = m\angle SQP$. Table 4-2 shows the S/R version of this proof.

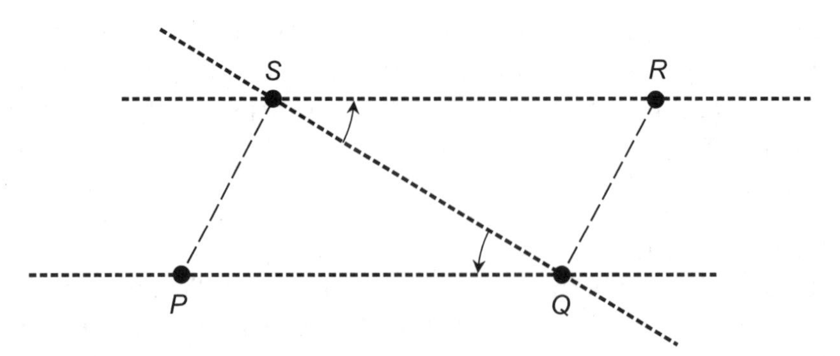

FIGURE 4-14 • Angles at vertices of triangles in a polygon. Illustration for Problem 4-6 and its solution.

TABLE 4-2 An S/R version of the proof demonstrated in the solution of Problem 4-6 and Fig. 4-14.

Statement	Reason
$\Delta SPQ \equiv \Delta QRS$	This is the theorem resulting from the solution of Problem 4-5.
Counterclockwise measures of corresponding angles in ΔQRS and ΔSPQ are equal.	This comes from the definition of direct congruence.
$\angle QSR$ and $\angle SQP$ are corresponding angles in ΔQRS and ΔSPQ, as we proceed in the same direction around both triangles.	This is evident from examination of the problem.
$m\angle QSR$ and $m\angle SQP$ are defined counterclockwise.	This is evident from the statement of the problem.
$m\angle QSR = m\angle SQP$	This comes from information derived in the preceding steps, and from the definition of direct congruence.

PROBLEM 4-7

In the scenario described by Problem 4-6, consider a point *T* on line *PQ* such that point *Q* lies between *P* and *T*. Let *U* be a point on line *SR* such that point *S* lies between *U* and *R*. Suppose that, in addition to the other conditions stated in Problem 4-6, we know that the following two statements hold true:

- **Line segment *QT* has the same length as line segment *US***
- **Line segment *UQ* has the same length as line segment *ST***

Prove that $m\angle USQ = m\angle TQS$.

SOLUTION

Work out this proof for yourself! If you have trouble, you might find the following hints helpful:

- **Construct triangles ΔUQS and ΔTSQ**
- **Show that ΔUQS and ΔTSQ are directly congruent**
- **Use Fig. 4-15 as a visual aid**

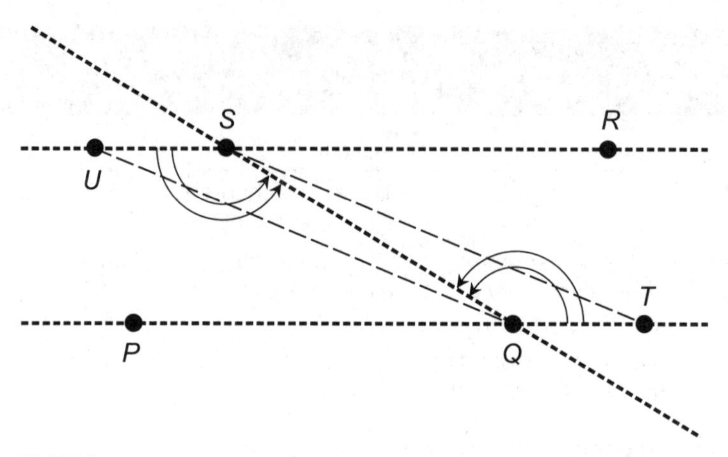

FIGURE 4-15 • Alternate interior angles. Illustration for Problem 4-7 and its solution.

TIP ***If you can't finish this proof, accept the proposition on faith for now, and try the proof again tomorrow.***

Alternate Interior Angles

Here's a new definition. Let *P*, *Q*, *R*, and *S* be distinct points, all of which lie in the same plane. Consider two distinct lines *PQ* and *SR*, both of which intersect a transversal line *SQ*. Let *T* be a point on line *PQ* such that *Q* lies between *P* and *T*. Let *U* be a point on line *SR* such that *S* lies between *U* and *R*. Then $\angle QSR$ and $\angle SQP$ constitute a pair of *alternate interior angles* formed by line *PQ*, line *SR*, and line *SQ*. Figure 4-14 shows these angles as arrowed arcs. In addition, $\angle USQ$ and $\angle TQS$ constitute a second pair of alternate interior angles formed by line *PQ*, line *SR*, and line *SQ*. Figure 4-15 shows these angles as arrowed double arcs.

PROBLEM 4-8

Prove that when two parallel lines *L* and *M* both intersect a transversal line *T*, pairs of alternate interior angles have equal measure. Let's call this the *AIA Theorem*.

SOLUTION

Try this for yourself, taking advantage of an additional postulate: "Given two parallel lines, both of which are transversal to two other parallel lines,

the pairs of corresponding parallel line segments between intersection points have equal lengths." Here's a hint: Take advantage of the solutions to Problems 4-6 and 4-7.

PROBLEM 4-9

Prove that the sum of the measures of the three interior angles of a triangle equals the measure of a straight angle.

SOLUTION

Consider an arbitrary triangle ΔPQR, with vertices at points P, Q, and R, expressed counterclockwise in that order. Let $x = m\angle RQP$, let $y = m\angle PRQ$, and let $z = m\angle QPR$. Figure 4-16A illustrates an example.

Choose a point S somewhere outside ΔPQR, such that line RS lies parallel to line PQ. Then choose a point T on line RS, such that R lies between T and S. We now have two parallel lines PQ and RS crossed by two different transversals PR and QR, as shown in Fig. 4-16B.

Let's consider $\angle QRS$, and call its measure x^*. Also consider $\angle TRP$, and call its measure z^*. Notice that $\angle RQP$ and $\angle QRS$, whose measures equal x and x^*, respectively, constitute alternate interior angles defined by the transversal line QR. Also notice that $\angle QPR$ and $\angle TRP$, whose measures are z and z^*, respectively, constitute alternate interior angles, defined by the transversal line PR. According to the AIA Theorem, it follows that $x = x^*$, and also that $z = z^*$.

From the geometry of this situation, we can see that the sum $z^* + y + x^*$ adds up to an angle whose measure is a straight angle (also called a 180° angle). In other words, $z^* + y + x^* = 180°$. The straight angle in this case is $\angle TRS$, formed by the collinear points T, R, and S. It follows that that $z + y + x = 180°$. (You'll get a chance to provide the reason in Quiz Question 1.) Notice that x, y, and z equal the measures of the interior angles of ΔPQR. Therefore, the sum of the measures of the interior angles of ΔPQR equals the measure of a straight angle. Table 4-3 shows an S/R version of this proof.

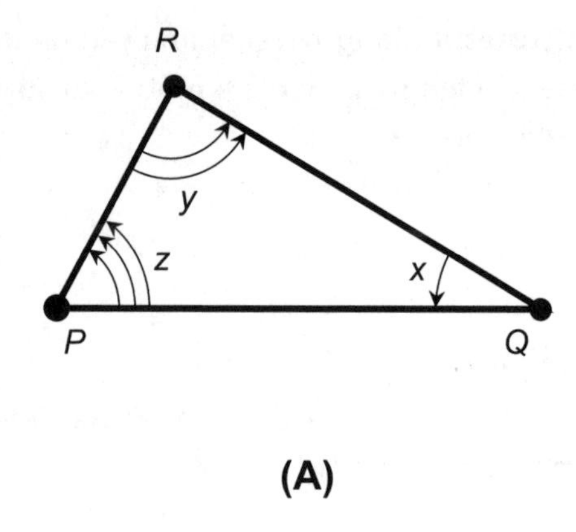

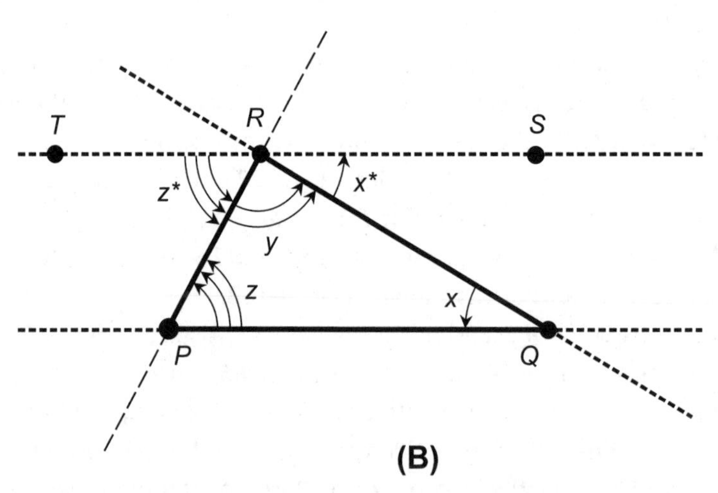

FIGURE 4-16 • At A, interior angles of a triangle. At B, construction of parallel lines and transversals. Illustration for Problem 4-9 and its solution.

PROBLEM 4-10

Imagine an isosceles triangle ΔPQR defined by three distinct points *P*, *Q*, and *R*, expressed counterclockwise in that order. Let *p* be the length of line segment *QR*. Let *q* be the length of line segment *RP*. Let *r* be the length of line segment *PQ*. Suppose that $q = r$. Let *S* be the point at the center of line segment *QR*, such that line segment *QS* has the same length as line segment *RS*. Line segment *PS* divides ΔPQR into two triangles, ΔPQS and ΔPSR. Prove that ΔPQS and ΔPSR are inversely congruent.

TABLE 4-3 An S/R version of the proof demonstrated in the solution of Problem 4-9 and Fig. 4-16.

Statement	Reason
Consider a triangle ΔPQR.	We have to start somewhere.
Choose a point S such that line RS is parallel to line PQ.	We will use this point later.
Choose a point T on line RS, such that point R is between point T and point S.	We will use this point later.
Line PR and line QR are transversals to the parallel lines PQ and RS.	This is apparent from the definition of a transversal line.
Consider $\angle QRS$, and call its measure x^*. Also consider $\angle TRP$, and call its measure z^*.	We will use these later.
Angles $\angle RQP$ and $\angle QRS$ are alternate interior angles.	This is apparent from the definition of alternate interior angles.
Angles $\angle QPR$ and $\angle TRP$ are alternate interior angles.	This is apparent from the definition of alternate interior angles.
$x = x^*$	This follows from the AIA Theorem.
$z = z^*$	This follows from the AIA Theorem.
$z^* + y + x^* = 180°$	This is apparent from the geometry of the situation, and from the definition of an angular degree.
$z + y + x = 180°$	You'll get a chance to fill this in later.
The measures of the interior angles of ΔPQR add up to the measure of a straight angle.	This is because x, y, and z are the measures of the interior angles of ΔPQR.

SOLUTION

Let's work around the two triangles, ΔPQS and ΔPSR, in opposite directions. We can rename the second triangle ΔPRS. That way, we go counterclockwise around ΔPQS, and clockwise around ΔPRS.

Let q^* be the length of line segment PS. Let p^* be the length of line segment QS. Let p^{**} be the length of line segment RS. (If you wish, you can draw a diagram of this situation. The result should look something like Fig. 4-17.) We are told that $q = r$. We know that $p^* = p^{**}$. (You'll get a chance to provide the reason in Quiz Question 2 at the

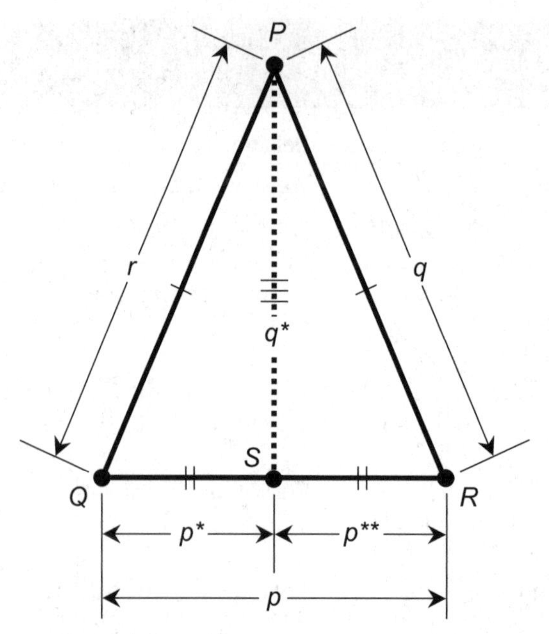

FIGURE 4-17 • Triangles within an isosceles triangle. Illustration for Problem 4-10 and its solution.

end of this chapter.) We also know that $q^* = q^*$! Let's state these three equations in order:

$$q = r$$

$$p^* = p^{**}$$

$$q^* = q^*$$

If we proceed counterclockwise around ΔPQS, we encounter sides of lengths r, p^*, and q^* in that order. If we proceed clockwise around ΔPRS, we encounter sides of lengths q, p^{**}, and q^* in that order. In this second case, we can substitute r for q because $q = r$, we can substitute p^* for p^{**} because $p^* = p^{**}$, and we know $q^* = q^*$. Therefore, if we proceed clockwise around ΔPRS, we encounter sides of lengths r, p^*, and q^* in that order. These are the same lengths, in the same order, as the lengths of the sides we encounter when we go counterclockwise around ΔPQS. By definition, the two triangles are inversely congruent. Table 4-4 shows an S/R version of this proof.

TABLE 4-4 An S/R version of the proof demonstrated in the solution of Problem 4-10 and Fig. 4-17.

Statement	Reason
Let q^* be the length of line segment PS.	We need to call it something!
Let p^* be the length of line segment QS.	We need to call it something!
Let p^{**} be the length of line segment RS.	We need to call it something!
$q = r$	We are told this.
$p^* = p^{**}$	You'll get a chance to fill this in later.
$q^* = q^*$	This is trivial. Anything is equal to itself.
Counterclockwise around ΔPQS, we encounter sides of lengths r, p^*, and q^* in that order.	This is evident from the geometry of the situation.
Clockwise around ΔPRS, we encounter sides of lengths q, p^{**}, and q^* in that order.	This is evident from the geometry of the situation.
Clockwise around ΔPRS, we encounter sides of lengths r, p^*, and q^* in that order.	This follows from substituting r for q and p^* for p^{**} in the preceding statement.
$\Delta PQS \equiv- \Delta PSR$	This follows from the statements in the first and third lines above this line, and from the definition of inverse congruence for triangles.

PROBLEM 4-11

Imagine a *regular hexagon*: a geometric figure with six vertices, six straight sides of identical length that connect adjacent pairs of vertices, and six interior angles of identical measure. As we go counterclockwise around the hexagon, let's name the vertices *P, Q, R, S, T*, and *U* in that order. Let *m* be the length of each side. Let *x* be the measure of each interior angle. Prove that $\Delta PQR \equiv \Delta STU$.

SOLUTION

If you wish, you can draw a diagram. The result should look something like Fig. 4-18. In ΔPQR, consider line segment PQ and line segment QR. Both of these sides have length m because they are sides of the hexagon, and

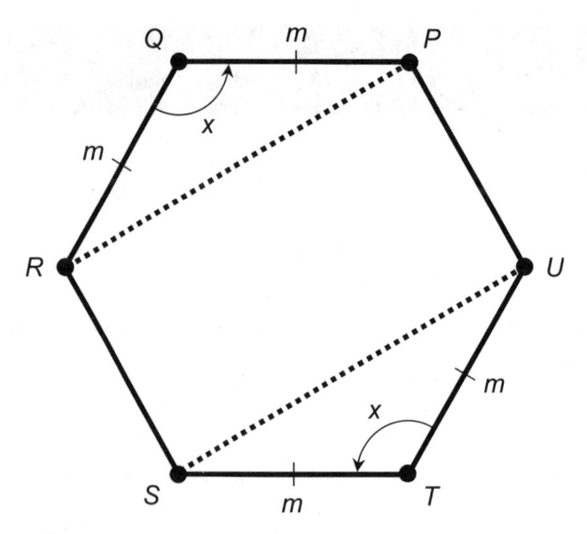

FIGURE 4-18 • Triangles within a regular hexagon. Illustration for Problem 4-11 and its solution.

we've been told that all the sides of the hexagon have length m. Consider the included angle between these two sides. It has measure x because it is an interior angle of the hexagon, and we've been told that all the interior angles of the hexagon have measure x. Proceeding counterclockwise from point P around ΔPQR, we encounter a side of length m, then an angle of measure x, and then a side of length m.

In ΔSTU, consider line segment ST and line segment TU. Both of these sides have length m because they are sides of the hexagon, and we've been told that all the sides of the hexagon have length m. Consider the included angle between these two sides. It has measure x because it is an interior angle of the hexagon, and we've been told that all the interior angles of the hexagon have measure x. Proceeding counterclockwise from point S around ΔSTU, we encounter a side of length m, then an angle of measure x, and then a side of length m. These side lengths and included angle measure are the same, and occur in the same order, as the corresponding side lengths and included angle measure in ΔPQR. Therefore, it follows that $\Delta PQR \equiv \Delta STU$. (You'll get a chance to provide the reason for this final step in Quiz Question 3.) Table 4-5 shows an S/R version of this proof.

TABLE 4-5 An S/R version of the proof demonstrated in the solution of Problem 4-11 and Fig. 4-18.

Statement	Reason
In ΔPQR, line segments PQ and QR both have length m.	Line segments PQ and QR are sides of the hexagon, and we are told that all sides of the hexagon have length m.
In ΔPQR, the included angle between the adjacent sides of length m has measure x.	This included angle is an interior angle of the hexagon, and we are told that all interior angles of the hexagon have measure x.
Counterclockwise from point P around ΔPQR, we encounter a side of length m, then an angle of measure x, and finally a side of length m.	This is evident from the geometry of the situation.
In ΔSTU, line segments ST and TU both have length m.	Line segments ST and TU are sides of the hexagon, and we are told that all sides of the hexagon have length m.
In ΔSTU, the included angle between the adjacent sides of length m has measure x.	This included angle is an interior angle of the hexagon, and we are told that all interior angles of the hexagon have measure x.
Counterclockwise from point S around ΔSTU, we encounter a side of length m, then an angle of measure x, and finally a side of length m.	This is evident from the geometry of the situation.
ΔPQR and ΔSTU have pairs of corresponding sides of identical lengths, with included angles of identical measures.	This is evident from the geometry of the situation.
$\Delta PQR \equiv \Delta STU$	You'll get a chance to fill this in later.

PROBLEM 4-12

Consider four distinct points *P*, *Q*, *R*, and *S*, all of which lie in the same plane. Suppose that lines *PQ* and *RS* are parallel. Suppose also that line segment *PQ* has the same length as line segment *RS*. Imagine two transversal lines *SQ* and *PR*, both of which cross lines *PQ* and *RS*, and which intersect at a point *T*. Prove that $\Delta PQT \equiv \Delta RST$.

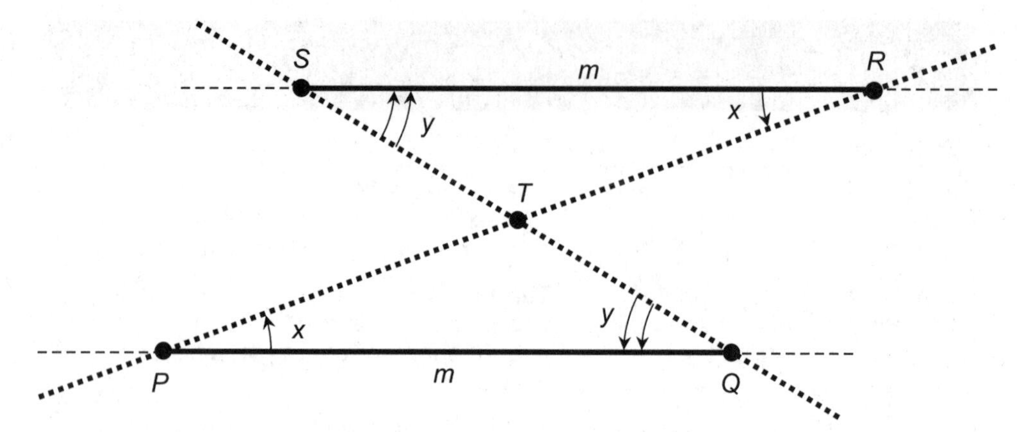

FIGURE 4-19 • Transversals through parallel lines. Illustration for Problem 4-12 and its solution.

SOLUTION

If you wish, you can draw a diagram. The result should look something like Fig. 4-19. Transversal line *PR* crosses both parallel lines *PQ* and *RS*. These three lines create a pair of alternate interior angles, $\angle SRT$ and $\angle QPT$. These two angles have equal measure. (You'll get a chance to provide the reason in Quiz Question 6.) Let's call this angle measure *x*.

Let *m* be the length of line segments *PQ* and *RS*. We can use *m* for both, because we've been told that they have equal lengths. Transversal line *SQ* crosses both parallel lines *PQ* and *RS*. These three lines form a pair of alternate interior angles, $\angle TQP$ and $\angle TSR$, which have equal measure. (You'll get a chance to provide the reason in Quiz Question 6.) Let's call this angle measure *y*.

Proceeding from point *P* counterclockwise around ΔPQT, we encounter first an angle of measure *x*, then a side of length *m*, and then an angle of measure *y*. Proceeding from point *R* counterclockwise around ΔRST, we encounter first an angle of measure *x*, then a side of length *m*, and then an angle of measure *y*. According to the ASA axiom, therefore, $\Delta PQT \equiv \Delta RST$. Table 4-6 shows an S/R version of this proof.

TABLE 4-6 An S/R version of the proof demonstrated in the solution of Problem 4-12 and Fig. 4-19.

Statement	Reason
Transversal line *PR* crosses lines *PQ* and *RS*.	This is evident from the geometry of the situation.
The two angles $\angle SRT$ and $\angle QPT$ are alternate interior angles within parallel lines.	This is evident from the geometry of the situation.
$\angle SRT$ and $\angle QPT$ have equal measure.	You'll get a chance to fill this in later.
Let x be the measure of $\angle SRT$ and $\angle QPT$.	We have to call it something!
Let m be the length of line segments *PQ* and *RS*.	We're told that their lengths are equal, and we have to call them something!
Transversal line *SQ* crosses lines *PQ* and *RS*.	This is evident from the geometry of the situation.
The two angles $\angle TQP$ and $\angle TSR$ are alternate interior angles within parallel lines.	This is evident from the geometry of the situation.
$\angle TQP$ and $\angle TSR$ have equal measure.	You'll get a chance to fill this in later.
Let y be the measure of $\angle TQP$ and $\angle TSR$.	We have to call it something!
Counterclockwise around ΔPQT or ΔRST, we encounter an angle of measure x, a side of length m, and an angle of measure y, in that order.	This is evident from the geometry of the situation.
$\Delta PQT \equiv \Delta RST$	This follows from the ASA axiom.

PROBLEM 4-13

Consider four mutually distinct points *P*, *Q*, *R*, and *S*, all of which lie in the same plane. Suppose that lines *PQ* and *RS* are parallel. Suppose also that $\angle PSR$ and $\angle RQP$ are right angles. Imagine a transversal line *PR* that crosses lines *PQ* and *RS*. Prove that $\Delta RPQ \equiv \Delta PRS$.

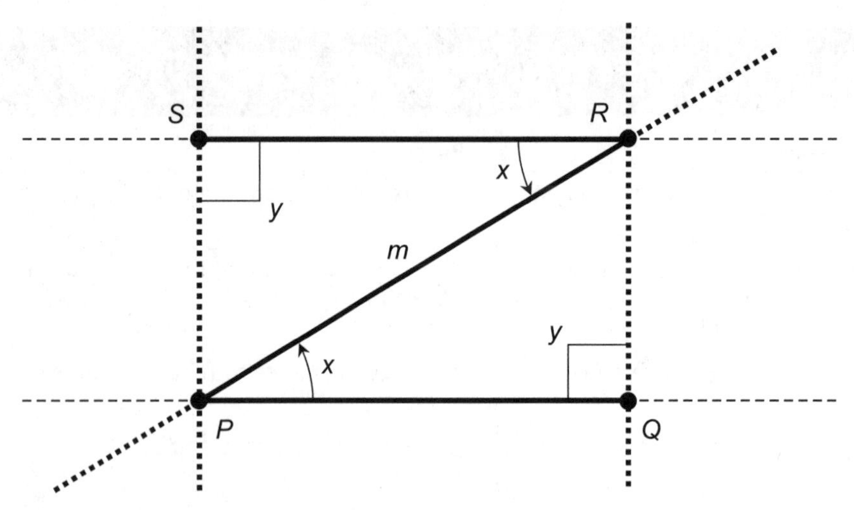

FIGURE 4-20 • Transversal through parallel lines, also forming the diagonal of a rectangle. Illustration for Problem 4-13 and its solution.

SOLUTION

If you wish, you can draw a diagram. The result should look something like Fig. 4-20. Let *m* be the length of line segment *RP*, which is the same as line segment *PR*.

A pair of alternate interior angles is formed by the transversal line *PR*, line *PQ*, and line *RS*. These angles are $\angle SRP$ and $\angle QPR$. According to the AIA Theorem, they have equal measure. Let *x* be this measure.

We've been told that $\angle PSR$ and $\angle RQP$ are right angles. Therefore, they have equal measure. (You'll get a chance to provide the reason in Quiz Question 8.) Let's call this angle measure *y*.

Proceeding from point *R* counterclockwise around ΔRPQ, we encounter first a side of length *m*, then an angle of measure *x*, and then an angle of measure *y*. Proceeding from point *P* counterclockwise around ΔPRS, we encounter first a side of length *m*, then an angle of measure *x*, and then an angle of measure *y*. According to the SAA axiom, therefore, $\Delta RPQ \equiv \Delta PRS$. Table 4-7 shows an S/R version of this proof.

TABLE 4-7 An S/R version of the proof demonstrated in the solution of Problem 4-13 and Fig. 4-20.

Statement	Reason
The length of line segment *RP* equals the length of line segment *PR*.	It doesn't matter which direction we go when we define the length of a line segment.
Let m be the length of line segments *RP* and *PR*.	We have to call it something!
$\angle SRP$ and $\angle QPR$ are alternate interior angles.	This is apparent from the geometry of the situation.
$\text{m}\angle SRP = \text{m}\angle QPR$	This follows from the AIA Theorem.
Let x be the measure of $\angle SRP$ and $\angle QPR$.	We have to call it something!
$\angle PSR$ and $\angle RQP$ are right angles.	We are told this.
$\text{m}\angle PSR = \text{m}\angle RQP$	You'll get a chance to fill this in later.
Let y be the measure of $\angle PSR$ and $\angle RQP$.	We have to call it something!
Counterclockwise around ΔRPQ or ΔPRS, we encounter a side of length m, an angle of measure x, and an angle of measure y, in that order.	This is evident from the geometry of the situation.
$\Delta RPQ \equiv \Delta PRS$	This follows from the SAA axiom.

QUIZ

You may refer to the text in this chapter while taking this quiz. A good score is at least 8 correct. Answers are in the back of the book.

1. **Refer to the solution of Problem 4-9 and Table 4-3. At one point in this proof, you are told that you'll get a chance to provide a reason for the statement later. You now have that chance! Which of the following sentences provides the correct reason?**
   - A. This conclusion follows from the SAA axiom.
   - B. We can substitute z for z^* and x for x^* in the preceding equation, because $z = z^*$ and $x = x^*$.
   - C. This conclusion follows from the definition of a triangle.
   - D. This conclusion follows from the parallel axiom.

2. **Refer to the solution of Problem 4-10 and Table 4-4. At one point in this proof, you're told that you'll get a chance to provide a reason for the statement later. You now have that chance! Which of the following sentences provides the correct reason?**
   - A. These are the lengths of line segments *QS* and *RS*, respectively, and we've been told that these line segments have equal lengths.
   - B. This conclusion follows from the parallel axiom.
   - C. This conclusion follows from the definition of a perpendicular lines; in this case lines *QR* and *PS*.
   - D. They have to be equal. If they were not, line *PS* would not intersect line *QR*, but these two lines obviously do intersect.

3. **Refer to the solution of Problem 4-11 and Table 4-5. At one point in this proof, you're told that you'll get a chance to provide a reason for the statement later. You now have that chance! Which of the following sentences provides the correct reason?**
   - A. This conclusion follows from the ASA axiom.
   - B. This conclusion follows from the AAA axiom.
   - C. This conclusion follows from the SAS axiom.
   - D. This conclusion follows from the SAA axiom.

4. **Unless otherwise specified, we should imagine angles going**
   - A. in the clockwise sense.
   - B. in the counterclockwise sense.
   - C. from left to right.
   - D. from right to left.

5. **Consider three mutually distinct lines *L*, *M*, and *N*. Which, if any, of the following scenarios A, B, or C is impossible?**
   - A. Line *L* is perpendicular to line *M*, and line *L* is also perpendicular to line *N*.
   - B. Line *M* is perpendicular to line *L*, and line *M* is also perpendicular to line *N*.
   - C. Line *N* is perpendicular to line *L*, and line *N* is also perpendicular to line *M*.
   - D. All three scenarios A, B, and C are possible.

6. **Refer to the solution of Problem 4-12 and Table 4-6. At two points in this proof, you are told that you'll get a chance to provide a reason for the statement later. You now have that chance! The same sentence can be inserted in both places to complete the proof. Which of the following sentences is it?**
   A. This conclusion follows from the ASA axiom, and the fact that line *PQ* is parallel to line *RS*.
   B. This conclusion follows from the SAA axiom, and the fact that line *PQ* is parallel to line *RS*.
   C. This conclusion follows from the definition of parallel lines, which tells us that line *PQ* is parallel to line *RS*.
   D. This conclusion follows from the AIA Theorem, and the fact that line *PQ* is parallel to line *RS*.

7. **What is the distinction between directly congruent triangles and inversely congruent triangles?**
   A. When two triangles are directly congruent, we never have to rotate one of them to make it coincide with the other; but when two triangles are inversely congruent, we usually have to rotate one of them to make it coincide with the other.
   B. When two triangles are directly congruent, we usually have to rotate one of them to make it coincide with the other; but when two triangles are inversely congruent, we never have to rotate one of them to make it coincide with the other.
   C. When two triangles are directly congruent, we never have to flip one of them over to make it coincide with the other; but when two triangles are inversely congruent, we usually have to flip one of them over to make it coincide with the other.
   D. When two triangles are directly congruent, we usually have to flip one of them over to make it coincide with the other; but when two triangles are inversely congruent, we never have to flip one of them over to make it coincide with the other.

8. **Refer to the solution of Problem 4-13 and Table 4-7. At one point in this proof, you are told that you'll get a chance to provide a reason for the statement later. You now have that chance! Which of the following sentences provides the correct reason?**
   A. This conclusion follows from the SAA axiom.
   B. This conclusion follows from the right angle axiom.
   C. This conclusion follows from the two-point axiom.
   D. This conclusion follows from the parallel axiom.

9. **In Fig. 4-21, suppose that all three lines intersect at a common point *T*. Suppose that $\angle QTR$ is a right angle. It follows that line *PQ* is perpendicular to line *RS* from**
   A. the two-point axiom.
   B. the right-angle axiom.
   C. the extension axiom.
   D. the definition of perpendicular lines.

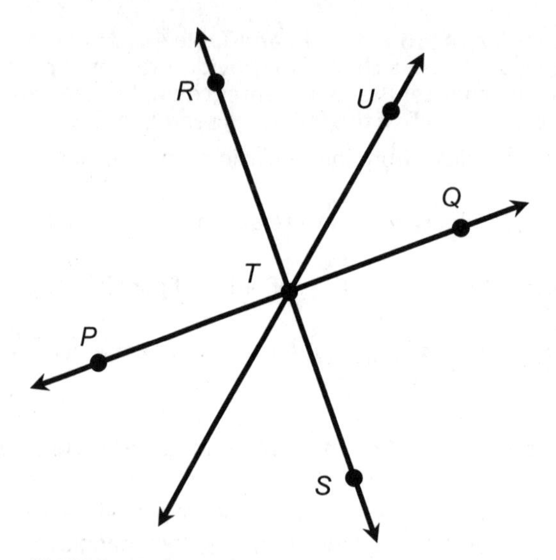

FIGURE 4-21 • Multiple lines intersecting at a common point. Illustration for Quiz Questions 9 and 10.

10. In Fig. 4-21, suppose that all three lines intersect at a common point *T*. Suppose that $\angle QTR$ is a right angle. It follows that $\angle QTU$ and $\angle UTR$

A. have equal measure.
B. are complementary.
C. are supplementary.
D. are alternate interior angles.

chapter 5

Fallacies, Paradoxes, and Revelations

In this chapter we'll examine some classical *fallacies*—violations or misapplications of the laws of reason. We'll also look at a few *paradoxes*, which are incredible results that arise from seemingly sound arguments. We'll conclude by looking at two of the most famous—and astounding—results ever obtained in logic.

CHAPTER OBJECTIVES

In this chapter, you will

- Avoid improper assumptions of probability.
- Identify common logical fallacies and learn to avoid them.
- Use inductive reasoning to strengthen arguments.
- Have fun with paradoxes, and try to resolve them.
- Understand Russell's paradox and Gödel's Incompleteness Theorem.

The Probability Fallacy

We say that a statement holds true when we've seen it or deduced it. If we believe that something is true or has taken place but we aren't sure, we're tempted to call it "likely." We would do well to resist such temptation.

Belief

When people formulate a theory, they often say that something "probably" happened in the distant past, or that something "might" exist somewhere, as yet undiscovered, at this moment. Have you ever heard that extraterrestrial life "probably" exists in our galaxy? Such a statement has no meaning. Either extraterrestrial life exists in our galaxy, or else it does not. The fact that we *don't know* (yet!) has no effect on this dichotomy.

If you say, "I think that the universe began with an explosion," you merely state that you believe it. You don't establish its truth, or even its "likelihood." If you declare outright, "The universe began with an explosion!" you state a theory, not a proven fact. If you say, "The universe probably started with an explosion," you insinuate that multiple pasts existed, and the universe had an explosive origin in more than half of them. In so doing, you commit the *probability fallacy* (PF) by injecting probability into a scenario where it has no place.

Whatever is, is. Whatever is not, is not. Whatever was, was. Whatever was not, was not. Either the universe started with an explosion, or else it did not. Either life exists on some other world in this galaxy, or else it does not.

Parallel Worlds, Fuzzy Worlds

If we say that the "probability" of life existing elsewhere in the cosmos is 20%, we say in effect, "Out of n observed universes, where n is some large natural number, $0.2n$ universes have been found to have extraterrestrial life." That doesn't mean anything to those of us who have seen only one universe!

We should put in a word of defense concerning forecasts such as "The probability of measurable precipitation tomorrow is 20%," or "The probability is 50% that John Doe, a cancer patient, will stay alive for more than 12 months." Statements such as these constitute hypothetical forecasts based on past observations of large numbers of similar cases. We can rephrase the weather forecast to say "According to historical data, when weather conditions have been as they are today, measurable precipitation has occurred the next day in 20% of the cases." We can rephrase the cancer situation to say "Of a large number of past cancer cases similar to that of John Doe, 50% of the patients lived beyond 12 months." These interpretations lend themselves to easy understanding. Therefore, these forecasters do not commit the PF. I, for one, would take a dim view of a meteorologist who said, "Either it will rain tomorrow, or else it will not," or a doctor who said, "Either your dad will stay alive for more than 12 months, or else he will not."

Still Struggling

Theories exist involving so-called *fuzzy truth*, in which some things "sort of happen." *Fuzzy logic* involves degrees of truth that range from completely false, through partially false, neutral, partially true, and totally true. Instead of only two values such as 0 (for falsity) and 1 (for truth), values can range along a smooth, unbroken interval or *continuum* from 0 to 1. In some cases the continuum has other limits, such as −1 to +1.

We Must Observe

We can assign probability values according to the results of observations, although we can also define probability on the basis of theory alone. When we abuse the notion of probability, seemingly sound reasoning can spawn absurd conclusions. Business people and politicians engage in this sort of misbehavior all the time, especially when they want to get someone to do something that will cause somebody else to get rich. Keep your "PF radar" switched on as you navigate your way through business and politics!

If you come across a situation where an author says that something "probably happened," "is probably true," "is likely to take place," or "is not likely to happen," think of it as a way for the author to state a belief or suspicion. Such a sentiment can have a basis in scientific observation or theory; it can have a basis in actual experience; it can have a basis in superstition. However, if we want to make sense out of this universe, we will do best if we adopt a stubborn "show me" attitude when we define reality.

Weak and Flawed Reasoning

When we break a rule of logic in a mathematical system, any derived result in that system becomes suspect. We can sometimes employ faulty logic to reach a valid conclusion that happens to be true, but a flaw in reasoning often results in a mistaken conclusion. We run the greatest risk of getting into trouble when the inaccurate conclusion offers intuitive appeal—it "seems true"—and the error reveals its existence only when we encounter a counterexample or a contradiction.

"Proof" by Example

The use of specific examples, even a great many of them, to prove general statements constitutes one of the most common fallacies in all of "illogic." The fact that we can sometimes get away with this fallacy makes it all the more perilous. Consider the following statement:

- Some rational numbers are integers.

Let R represent the predicate "is a rational number," Z represent the predicate "is an integer," and x represent a variable from the set of all numbers. We can symbolize the above statement as:

$$(\exists x)\ \mathrm{R}x \wedge \mathrm{Z}x$$

We can prove this statement by demonstrating that it works for a single rational number, such as 35/5. That rational number equals 7, which constitutes an integer. Once we've shown that the statement holds true in one case, we've satisfied the existential quantifier "For some," which also means "There exists at least one."

Now imagine that we find ourselves confronted with the following, more powerful proposition:

- All integers are rational numbers.

When we put this sentence into symbolic form, we get an expression that contains a universal quantifier:

$$(\forall x)\ \mathrm{Z}x \rightarrow \mathrm{R}x$$

We can find plenty of specific cases for which this proposition holds true. We can take an integer, such as 35, and then divide it by 1, getting 35 as the quotient, and then claim that the original integer is equal to this quotient. That's trivial:

$$35 = 35/1$$

This is a rational number, because it's a quotient in which the numerator is an integer and the denominator is a nonzero natural number. We can do the same exercise with many integers, always putting 1 in the denominator, generating countless examples:

$$40 = 40/1$$

$$-45 = -45/1$$

$$260 = 260/1$$

$$\ldots$$

and so on, *ad infinitum*

Confident in our example-showing skills, we can arrange the set Z of all integers as a list:

$$Z = \{0, 1, -1, 2, -2, 3, -3, \ldots\}$$

(Note that we represent the set of integers with an uppercase italic Z, whereas we represent the predicate "is an integer" with uppercase nonitalic Z.) We can rewrite the set list so that every element comprises a quotient with 1 in its denominator, making every element obviously rational, as follows:

$$Z = \{0/1, -1/1, 1/1, 2/1, -2/1, 3/1, -3/1, \ldots\}$$

This exercise demonstrates the truth of our proposition for as many examples as we have the time and inclination to list. As the length of our list grows, we find it increasingly tempting to suppose that the proposition holds true for *all* integers. Almost any reasonable person would come to the conclusion, after testing for a few specific integers, that the proposition must hold true in every possible case. But specific examples, no matter how numerous, do not *rigorously* prove the proposition and thereby allow us to call it a theorem.

We should avoid the use of examples—even millions of them—to prove general propositions that contain universal quantifiers. In this particular instance, we have at least two options.

- We can force the element in question to act as a variable such as x (and not a constant, such as 35) and use deductive logic, armed with the laws of arithmetic, to derive the desired conclusion.

- We can apply a technique known as *mathematical induction* to the set of integers after arranging it as an "implied list," as we have done with Z above. We'll learn how to use mathematical induction in Chap. 6.

Begging the Question

You will sometimes hear or read "proofs" that do nothing beyond assuming the truth of a proposition straightaway. If you've a lawyer, you will recognize this fallacy the instant you hear it or read it. Have you ever pointed out a huge flaw of this nature to your opponent in an argument or debate? Have you ever tried to get away with it in your own arguments? We call this fallacy *begging the question*.

When you beg a question, you don't logically prove anything, whether the proposition is true or not. You prove only a propositional-calculus triviality known as a *tautology*:

$$X \rightarrow X$$

Here's an example of begging the question. Suppose the temperature is 40 degrees below zero Celsius, and the wind is gusting to 100 kilometers per hour (km/h). You state this fact and then conclude, "It is cold and windy today!" Here's another example. Suppose John Doe hit 100 home runs during last year's baseball season. Your friend informs you of this fact and then says, "John Doe hit many home runs last year!" Neither of these conclusions prove anything. They constitute rephrasing or restatement of obvious truths, and nothing more.

TIP ***People sometimes beg the question in ways subtle enough to quality as perverse art forms! "It imperils the population to have motor vehicles moving at high speeds in residential areas. Therefore, if we allow people to drive cars and trucks on the streets of our cities at unlimited speeds, it presents a danger to the community." This eloquent mass of verbiage merely states the same thing twice. It proves nothing whatsoever.***

Sometimes, begging the question takes a roundabout form leading from the premise through a labyrinth of valid logic, but ultimately arriving back at the original premise. After that exercise, the fallacy-monger claims to have proven something profound, when in fact she has purveyed a great deal of useless information.

Hasty Generalization

When we commit the fallacy of *hasty generalization*, we assign a certain characteristic to something as a whole, based on examination of the wrong data, incomplete data, or data that's both wrong and incomplete.

Suppose that every time you ask people for favors when you wash your laundry, they refuse. What if this event occurs 12 times in a row? Let L represent the statement, "You wash your laundry." Let F represent the statement, "You ask someone to do you a favor." Let T represent "The person does you the favor you asked for." You find that the event $\neg T$ repeatedly follows the event $(L \wedge F)$. In your frustration, you use this set of unfortunate experiences to "prove" that

$$(L \wedge F) \rightarrow \neg T$$

This "proof" does not constitute valid logical reasoning. The fact that something has happened in numerous instances doesn't mean that it happens in all instances, or even that it will happen in the very next instance.

Misuse of Context

A word can have two or more different meanings, depending on the context in which it occurs. But in a valid logical argument, we must never alter the intended meaning of a word in the course of the discussion. Such carelessness (or deviousness) can result in misleading, absurd, or nonsensical conclusions.

Consider the statement "You can't keep cows in a pen that has run out of ink." Here, the word "pen" refers to an enclosure for animals in the first instance, and a writing instrument in the second instance. This statement provides us with an extreme example of the sort of nonsense that can result from *misuse of context*.

Circumstance

Devious folks sometimes cobble together "arguments" in an effort to lead other people to come to a mistaken conclusion. The conclusion seems reasonable enough "at first thought," but no true logical proof actually takes place. When we commit this fallacy, we formulate an *argument by circumstantial evidence*.

We often encounter arguments by circumstantial evidence in criminal trials. A lawyer "sets up" witnesses by asking questions not directly related to the crime. For example, suppose that I have been accused of some misdemeanor, and the state puts me on trial. Several witnesses claim that they saw me in the vicinity of the place where the crime occurred. Other witnesses testify that I was away from home when the crime took place. Still other witnesses express the opinion that I'm a no-good, rotten son-of-a-buck. Even a dozen, or a hundred, or a thousand such testimonials do not *rigorously prove* that I committed the crime. Even if my guilt can be inferred beyond "reasonable" doubt, these testimonials do not provide a *mathematical* proof of my guilt.

Fallacies with Syllogisms

In a *syllogism*, we draw a conclusion based on two premises. You learned about syllogisms in Chap. 3. The first premise usually comprises an "or" statement (disjunctive syllogism) or an "if-then" statement (hypothetical syllogism). Syllogisms allow us to commit fallacies, sometimes without realizing that our "logic" contains any flaw. Consider the following disjunctive syllogism:

- Jill is in Florida or Jill is in New York. Jill is not in Florida. Therefore, Jill is in New York.

Let F represent the predicate "is in Florida," N represent "is in New York," and j represent the constant "Jill." We can symbolize the argument as follows:

$$Fj \vee Nj$$

$$\neg Fj$$

$$Nj$$

Here's an example of a hypothetical syllogism:

- Anyone who takes 100 sleeping pills all at once will die. Joe took 100 sleeping pills all at once. Therefore, Joe will die.

Let P represent the predicate "takes 100 sleeping pills all at once." Let D represent "will die." Let x be a logical variable, and let j represent the constant "Joe." Then symbolically, the argument looks like this:

$$(\forall x)\ Px \rightarrow Dx$$

$$Pj$$

$$Dj$$

We can refute (disagree with or disprove) one or the other of the premises in either of these syllogism examples, but in themselves, the arguments remain logically valid.

A crafty deceiver can "twist" a syllogism to create a fallacious argument. We can make such a logical error or do a deed of subterfuge if we *deny the antecedent*. Consider the following argument:

- If you commit a federal offense, you'll go to prison. You did not commit a federal offense. Therefore you will not go to prison.

Let F represent the predicate "commit(s) a federal offense." Let P represent "will go to prison." Let *y* represent the constant "you." The above argument appears as follows when we write it in symbolic form:

$$\begin{array}{l} Fy \rightarrow Py \\ \underline{\neg Fy \qquad\qquad\qquad} \\ \neg Py \end{array}$$

This syllogism does not constitute a logically valid argument. You can commit plenty of *nonfederal* crimes (acts that are illegal in, say, South Dakota but entirely legal in California) that will land you in jail if you get caught. Besides that, some entirely innocent people can end up in prison.

The foregoing fallacy can also occur if the original antecedent constitutes a negative statement. Consider this:

- If John was not near the grocery store last night, he must have been at home. John was indeed near the store last night. Therefore, he was not at home.

Let G represent the predicate "was near the grocery store last night." Let H represent "was at home." Let *j* represent the constant "John." Then symbolically, we have the following syllogism:

$$\begin{array}{l} \neg Gj \rightarrow Hj \\ \underline{Gj \qquad\qquad\qquad} \\ \neg Hj \end{array}$$

This argument does not constitute a valid line of reasoning. What if John's home sits right next door to the grocery store? Then most people would say that John is near the grocery store even when he's at home. But even in that case, the preceding argument contains a lack of clarity in the meaning of the word "near." You might think "near" means "within 100 meters of" while I might think it means "within a kilometer of" and our friend Mike thinks it means "within 100 kilometers of." Lawyers will recognize (and often exploit) this sort of trick!

Yet another fallacy can occur in disjunctive syllogisms. Consider the following dilemma and argument:

- Wanda must leave the country or get arrested for a crime of which she has been accused. Wanda has left the country. Therefore, Wanda will not get arrested.

Let L represent the predicate "must leave the country." Let A represent "will get arrested for a crime of which she has been accused." Let w represent the constant "Wanda." Then symbolically, our argument looks like this:

$\mathrm{L}w \vee \mathrm{A}w$

$\mathrm{L}w$

$\neg \mathrm{A}w$

We have a fallacy here. Wanda might get arrested even if she leaves the country. We have subtly confused the inclusive and exclusive forms disjunction to arrive at a conclusion that the premises do not logically support.

Fun with Silliness

We can take advantage of nonsensical subject-predicate combinations to demonstrate logical validity or invalidity. This tactic overcomes the human tendency to assign notions about everyday life to logical derivations. Consider the following argument:

- If the moon consists of Swiss cheese, then some ants eat chocolate. The moon is made of Swiss cheese. Therefore, some ants eat chocolate.

Let S represent the predicate "consists of Swiss cheese." Let A stand for "is an ant." Let C stand for the predicate "eats chocolate." Let m represent the constant "the moon," and let x represent a logical variable. We can symbolize the above argument as follows:

$\mathrm{S}m \rightarrow [(\exists x)\ \mathrm{A}x \wedge \mathrm{C}x]$

$\mathrm{S}m$

$(\exists x)\ \mathrm{A}x \wedge \mathrm{C}x$

We have a logically valid argument here, but suppose that we deny the antecedent. Then we can argue that because the moon does not consist of Swiss cheese, no ant will eat chocolate. That's fallacious. Hungry ants would swarm all over a warm piece of chocolate, regardless of the material composition of the moon.

Let's look at another syllogism that contains a disjunction, and that obviously constitutes a fallacy when we apply it to the "real world." Scrutinize the following sequence of nonsensical statements:

- Either Mars is inhabited by little green rabbits, or the sky appears blue as seen from the surface of the earth on a clear day. No little green rabbits dwell on Mars. Therefore, the sky appears blue as seen from the surface of the earth on a clear day.

Let R represent "is inhabited by little green rabbits." Let B represent "appears blue as seen from the surface of the earth on a clear day." Let *m* represent the constant "Mars." Let *s* represent the constant "the sky." Then we can symbolize the above argument as follows:

$$\begin{array}{l} \mathrm{R}m \vee \mathrm{B}s \\ \neg \mathrm{R}m \\ \hline \mathrm{B}s \end{array}$$

In this scenario, the first statement (a disjunction for this particular case) in a syllogism is always true. The logic holds valid, but the example proves nothing. If there were little green rabbits on Mars, the sky would nevertheless appear blue as seen from the surface of the earth on a clear day. By plugging in other predicates and constants, logic tricksters can convince people that one event causes, or correlates, with another event, when in fact no connection exists whatsoever.

Inductive Reasoning

When we use *inductive reasoning,* we attempt to show that some proposition holds true most of the time, or that we can reasonably expect that some event will occur. Inductive reasoning differs from the collection of logical principles that we learned in Chaps. 2 and 3, also known as *deductive logic*. While deductive logic constitutes a rigorous method of proof, inductive reasoning does not.

Logic tricksters occasionally portray inductive reasoning as if it can work as a rigorous deduction, fooling people into thinking that a flawed argument constitutes valid logic. To make things worse, the trickster can state a conclusion to the effect that something "is probably true" or "will probably occur" or "probably took place," thereby invoking the probability fallacy (PF) in addition to letting flimsy reasoning masquerade as valid logic.

Let's look at an example of what can happen when we combine inductive reasoning with the PF, generating an absurd conclusion. Suppose that the legal speed limit on a stretch of highway is 100 km/h. A police officer who needs to meet a weekly quota (for issuing tickets to "speeders," or people who drive too fast) interprets this speed limit to mean that you're a "speeder" if you drive at *100 km/h or more* (as opposed to *more than 100 km/h*). Imagine that you cruise in your new car along the highway at 99.6 km/h, and the police officer's radar equipment displays your speed digitally to the nearest kilometer per hour, rounding it off to 100 km/h. The officer sees this number and concludes that you're *probably* a "speeder." His reasoning, and your unfortunate fate, proceeds as follows:

- Given the radar reading, the probability that you're moving at 100 km/h or more equals exactly 50%, because your exact, true speed must be more than 99.5 km/h but less than 100.5 km/h.
- If we round off 50%, or 0.5, to the nearest whole digit, then that digit equals 1, or 100%, according to mathematical convention.
- The officer rounds the probability of 0.5 up to a probability of 1, which constitutes certainty.
- The conclusion: You are a "speeder."
- The officer issues you a citation.

In the officer's fallacy-tainted mind, a reading of 100 km/h on radar means that you're driving at 100 km/h or more if you drive at 99.5 km/h or more! Now you must convince the judge in the traffic court that the officer has misused inductive reasoning and committed the probability fallacy. Good luck!

PROBLEM 5-1

Imagine that someone makes the following statement and claims it as a mathematical theorem:

- **All rational numbers can be written as *terminating decimals*, that is, as decimal numbers where the digits after a certain point are all zeros.**

You claim that this cannot be a theorem because it simply isn't true. How many counterexamples must you find in order to show that this theorem is not true?

SOLUTION

You only have to find one counterexample to demonstrate that a claimed theorem is not true. In this case, you can cite the fact that 1/3 = 0.333 . . ., which is not a terminating decimal because the numeral 3 repeats without end. Yet, 1/3 is obviously a rational number according to the traditional definition of the term (a ratio of integers).

PROBLEM 5-2

What sort of fallacy exists in the following argument? Symbolize the argument, and then identify the fallacy. Note that a *polygon* comprises a geometric figure that lies entirely in a single plane, and that has three or more straight sides such that all adjacent pairs of sides intersect at their end points, and no two sides intersect except at their end points. A *triangle* is a polygon with three sides.

- **All triangles are polygons. Figure *S* is not a triangle. Therefore, figure *S* is not a polygon.**

SOLUTION

You and I both know from experience that there exist plenty of polygons besides triangles, such as squares, rectangles, trapezoids, pentagons, hexagons, heptagons, octagons, and so on. The above argument obviously contains a flaw. Let T represent the predicate "is a triangle," let P represent the predicate "is a polygon," let *x* represent a logical variable, and let *s* represent the constant "figure *S*." Now we can rearrange the argument so that it reads like this:

- **For all *x*, if *x* is a triangle, then *x* is a polygon. It is not true that *s* is a triangle. Therefore, it is not true that *s* is a polygon.**

Symbolically, we can write this argument as

$(\forall x)\ Tx \rightarrow Px$

$\underline{\neg Ts}$

$\neg Ps$

We have denied the antecedent. Note that we can "turn the argument around" and make it into a valid sequence of reasoning. Symbolically:

$(\forall x)\ Tx \rightarrow Px$

$\neg Ps$

$\neg Ts$

This translates to the following three sentences:

- **All triangles are polygons. Figure *S* is not a polygon. Therefore, figure *S* is not a triangle.**

Simple Paradoxes

Following are a few little tidbits that seem to defy logic. They show what can happen when we try to apply mathematical rigor to nonrigorous "real-world" facts and ideas.

A Wire around the Earth

We all know about the irrational number pi (π), which equals approximately 3.1416 and represents the number of diameters in the circumference of a circle or sphere. This constant applies without fail in Euclidean geometry; the size of the circle or sphere makes no difference.

We can derive an interesting—and quite counterintuitive—result from the formula for the circumference of a circle sphere based on its diameter. The formula is

$$c = \pi d = 2\pi r$$

where c represents the circumference, d represents the diameter, and r represents the radius, all expressed in the same units (such as meters).

Suppose that the earth were a smooth, perfectly round sphere, with no hills or mountains. Imagine a perfectly nonelastic wire wrapped snugly around the earth's equator. If we add 10 meters (10 m) to the length of this wire, and then prop it up all the way around the world so that it stands out equally far from the surface at every point, how far above the earth's surface will the wire stand? Assume the earth's circumference equals 40,000,000 m.

Most people think that the wire will stand out only a tiny distance from the surface of the earth if we make it 10 m longer. After all, that extra 10 m represents a wire-length increase of only 10 parts in 40,000,000, or 25 millionths of one percent! But in fact the wire will stand out approximately 1.59 m all the way around the earth. For those of you not familiar with the metric system, that's almost 5 ft 3 in.

PROBLEM 5-3

We don't have to content ourselves with showing that the above-mentioned "trick of illogic" works for a globe the size of the earth. We can extend the assertion to claim that if we add 10 m to the length of a wire that tightly girdles the circumference of any sphere, no matter how large, the lengthened wire will stand out the same distance from the surface: approximately 1.59 m. Prove this!

SOLUTION

Refer to Fig. 5-1. Suppose that the radius of the sphere, expressed in meters, is equal to *r*. Suppose that the circumference of the sphere, also expressed in meters, is equal to *c*. From the rules of geometry, we know that

$$c = 2\pi r$$

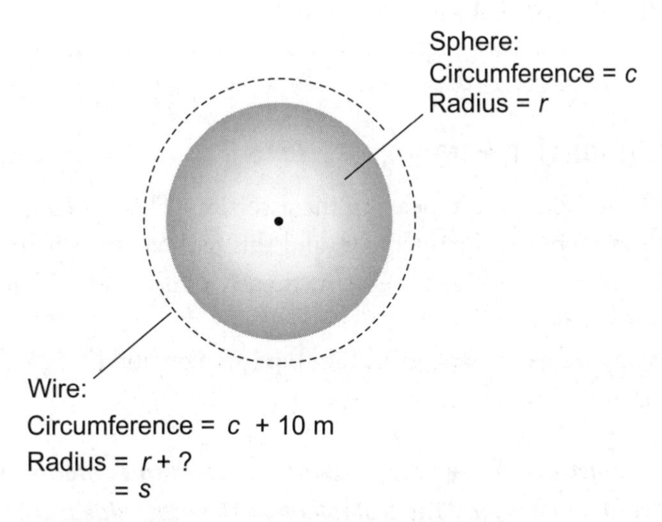

FIGURE 5-1 • If we increase the circumference of a circular wire loop around a planet by 10 m, by how much does the loop's radius increase?

Solving the equation for r gives us

$$r = c/(2\pi)$$

If we add 10 m to the length of a wire whose original length is equal to c (because it girdles the sphere around a circumference), then the lengthened wire lies in a circle whose radius s, expressed in meters, can be calculated as follows if we let $\pi = 3.14$:

$$\begin{aligned} s &= (c + 10)/(2\pi) \\ &= c/(2\pi) + 10/(2\pi) \\ &= c/(2\pi) + 10/(2 \times 3.14) \\ &= c/(2\pi) + 10/6.28 \\ &= c/(2\pi) + 1.59 \\ &= r + 1.59 \end{aligned}$$

We've just shown that the radius of the circle described by the lengthened wire is 1.59 m greater than the radius of the sphere. Therefore, the lengthened wire can be positioned so that it stands 1.59 m above the surface of the sphere, all the way around. We don't have to specify a value for r, because it doesn't make any difference!

Direct-Contradiction Paradox

A common logical paradox appears in the assertion "This statement is false." If we affirm the statement, then it's true, and this fact contradicts its own claim that it's false. If we deny the statement, then it's false, but it denies its own falsity, making it true.

The only way we can escape from the trap is to conclude that the statement has no truth value whatsoever.

TIP ***We can construct somewhat less blunt version of this paradox by taking an index card and writing "The statement on the other side is true" on one face. Then we can flip the card over and write "The statement on the other side is false."***

Who Shaves Hap?

Imagine that Hap is a barber in the town of Happyton. Hap shaves all the people, but only the people, in Happyton who do not shave themselves. Does Hap shave himself, or not?

Assume Hap shaves himself. We've just stated that he shaves only those people who do not shave themselves. Therefore, Hap does not shave himself. So Hap shaves himself and Hap does not shave himself—a contradiction. By *reductio ad absurdum*, Hap does not shave himself.

Unfortunately, the above conclusion leads us to the fact that Hap shaves himself, because he shaves all the people (including himself) that do not shave themselves. Again, we invoke *reductio ad absurdum*, deriving the fact that Hap shaves himself. We run in a circle of contradictions.

We have only one clear way out of this logical vortex: We must conclude that no such person as Hap can live in Happyton, or that the town of Happyton can't exist at all.

Arrow Paradox

Imagine that Robin Hood shoots an arrow towards a target. Consider a particular point in time (i.e., an instant with no duration but representing a definite clock reading) while the arrow flies. We can logically deduce that one or the other of the following statements must hold true at that instant:

- The arrow moves to the point in space where it is
- The arrow moves to a point in space where it is not

Let's suppose that the first statement holds true: The arrow moves to the point in space where it is. Upon reflection, we realize that this claim makes no sense. The arrow already is where it is! Nothing can *move* to a point that it already occupies, because that notion implies that the object can move and not move at the same time.

If we assume the truth of the second statement, we run into trouble again, but of a different sort. An arrow can move to a point where it *was not* a little while ago, but nothing can arrive a point where it *is not*. If Robin's arrow could do such a thing, it could simultaneously exist and not exist at all points in its flight.

Apparently, once Robin Hood releases an arrow from his bow, that arrow cannot travel through space. By extrapolation, we must conclude that nothing in the universe can move.

The Frog and the Wall

A familiar problem in mathematics involves the adding-up, or *summing*, of an *infinite sequence* in order to get a finite sum. The *frog-and-wall paradox* (or what at first seems to constitute a paradox) shows an example of how this counter-intuitive principle works.

Imagine a frog standing with her nose exactly 8 m away from a wall. Imagine that the frog jumps halfway to the wall, so that after the jump her nose is exactly 4 m from the wall. Suppose that she continues to jump toward the wall, each time getting halfway there. She'll never reach the wall if she jumps in this fashion, no matter how many times she jumps, even though she has only 8 m to travel at the outset. The frog will die before she gets to the wall, even though she'll almost reach it. Figure 5-2 illustrates the first few jumps in the sequence. No finite number of jumps will allow the frog to reach the wall. To attain that goal, she'd have to make an infinite number of jumps, an impossibility in the "real world."

We can represent this scenario as the following *infinite series*. (A *series* is the sum of the terms in a *sequence*.) Let's call it S, so that we have

$$S = 4 + 2 + 1 + 1/2 + 1/4 + 1/8 + \ldots$$

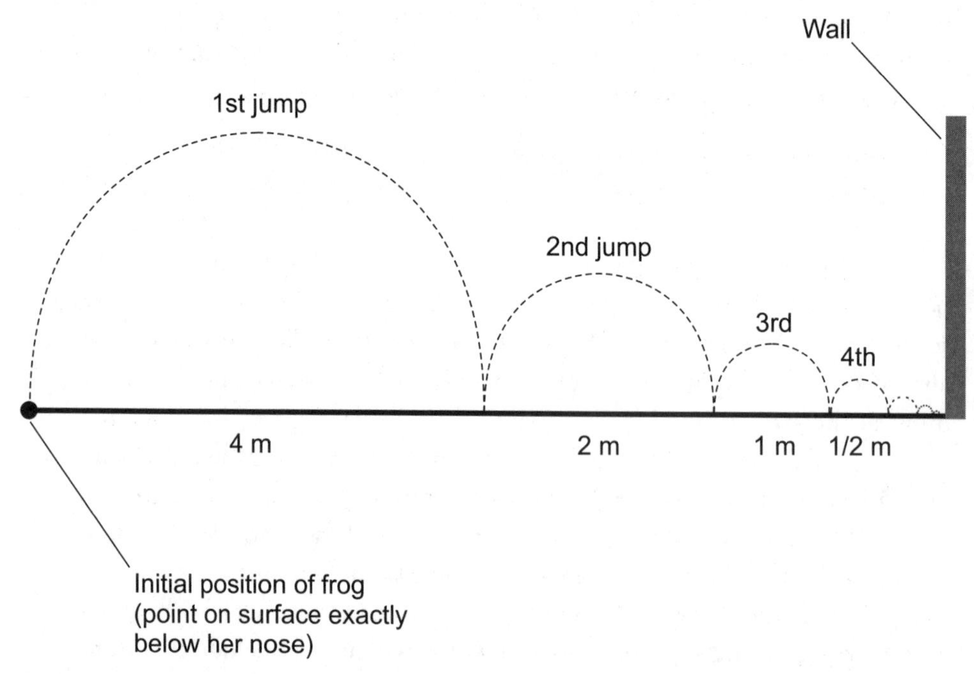

FIGURE 5-2 • A frog jumps towards a wall, getting halfway there each time.

If we keep cutting a number in half over and over *without end*, and sum up the total of all the values in the resulting infinite sequence, we will obtain twice the original number. In this particular scenario, that means we'll get $S = 8$.

How, you ask, can we actually add up the numbers in an infinitely long sequence? In the "real world," of course, we can't do it. We can approach the actual sum, but we can never quite get all the way there because we don't have enough time (i.e., an infinite amount of time) to add up an infinite number of numbers. But in the mathematical world, certain infinite series add up to finite numbers. The constraints of physical reality do not bind us in the universe of pure mathematics. When an infinite series has a finite sum, we call that series *convergent*.

A "real-world" frog cannot reach a wall by jumping halfway to it, over and over. But a mathematical frog can! It can happen in two ways. First, there exists an infinite "supply" of time, so an infinite number of jumps can take place. Another way around the problem comprises repeatedly halving the length of time in between jumps, say from four seconds (4 s) to 2 s, then to 1 s, then to 1/2 s, and so on. This exercise allows the frog to make infinitely many hops in a finite span of time—although, toward the end of the journey, she'll be hopping mighty fast!

PROBLEM 5-4

We can use the same sort of argument as the one in the frog-and-wall scenario to "prove" that if you drive a car at 80 km/h and try to pass another car ahead of you that's traveling at 50 km/h along the same road, you'll never catch, let alone pass, that car. How does this "proof" work? What's wrong with it?

SOLUTION

Figure 5-3 illustrates this situation. Drawing A shows the initial state of affairs. You, moving at 80 km/h, are located a certain distance d_0 (in meters) behind the car that you want to catch, which travels at 50 km/h. (For reference, define the "point" that a car occupies as the point on the road exactly below the point at the center of the car's forward bumper.) After a certain span of time, your car has traveled d_0 m, reaching the position previously occupied by the other car. But in that same period of time, that

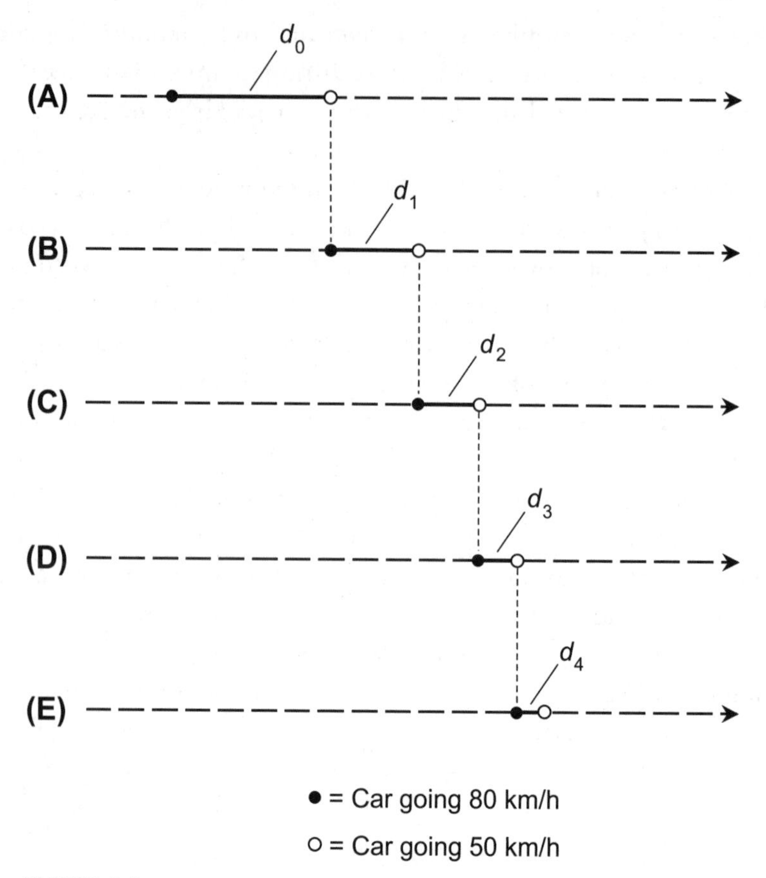

FIGURE 5-3 • Illustration for Problem 5-6. A fast car comes up behind a slower car. The distances d_0, d_1, d_2, d_3, d_4, . . . keep getting smaller.

other car has moved farther ahead by a distance of d_1 m, so it's in a new position d_1 m in front of your car (drawing B). The distance d_1 is less than the distance d_0. After a little more time, your car has traveled the distance d_1 m, but the other car has moved ahead yet a little more, and now occupies a point d_2 m in front of your car (drawing C). The same thing keeps taking place over and over. Drawings D and E show what happens in the next two time frames, as you travel the distance d_2 and then the distance d_3. At E, the other car leads yours by d_4 m. This "catch-if-catch-can" process goes on without end. Conclusion: Your car can never catch the leading car.

In the "real world," your car will catch and pass the slower car because the sequence $d_0, d_1, d_2, d_3, d_4, \ldots$ is such that its corresponding series, T, converges. We have

$$d_0 > d_1 > d_2 > d_3 > d_4 > \cdots$$

such that the infinitely long sum

$$T = d_0 + d_1 + d_2 + d_3 + d_4 \cdots$$

equals a finite, definable real number. The "trick" lies in the fact that the sum of the time intervals corresponding to the transitions of the distance intervals is finite, not infinite. You span the distances $d_0, d_1, d_2, d_3, d_4, \ldots$ at an ever-increasing rate as you race down the highway to pass that slow-poke in front of you. The rate at which you span the progressively smaller intervals "blows up." You in effect add up an infinite number of numbers, corresponding to smaller and smaller distance intervals, in a finite length of time.

A Geometry Trick

Here's a graphical puzzle that, at first glance, seems to defy the laws of Euclidean geometry. The trick distorts simple line drawings (ever so slightly!) to help a would-be deceiver come to an invalid conclusion. Figure 5-4 illustrates the scheme, and its resolution, in four stages.

We begin by drawing a square on a sheet of paper and then dividing it into 64 smaller squares. Then we draw dashed lines inside the main square as shown in Fig. 5-4A. We cut the paper along the dashed lines, obtaining two right triangles X and Z, each measuring 3-by-8 units, and two trapezoids, W and Y, consisting of 3-by-5 rectangles "stuck" to 2-by-5 right triangles. We then rearrange these four pieces to create a 13-by-5 rectangle as shown in Fig. 5-4B. This rectangle measures 13 units long by 5 units wide, so that its total area equals 13 × 5, or 65, square units—a full square unit larger than the area of the original square.

It appears as if, by merely cutting up a sheet of paper and rearranging the pieces, we've created a square unit of paper from nothing!

(A)

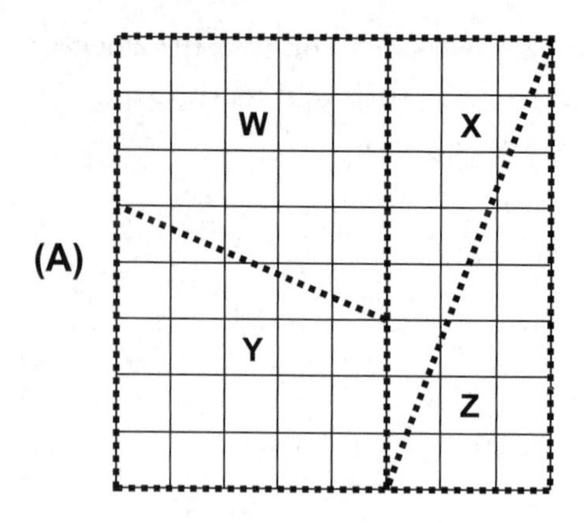

(B)

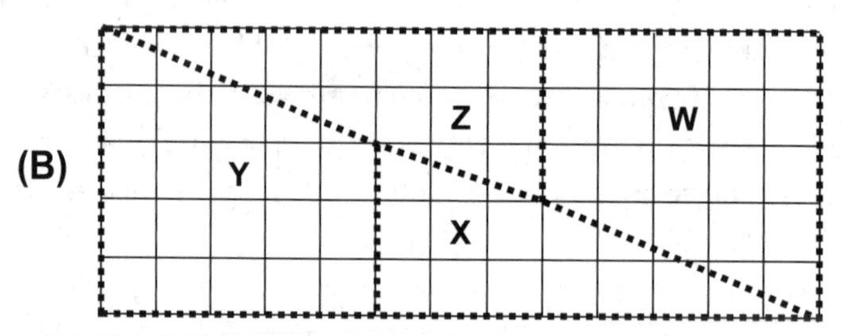

FIGURE 5-4 • At A, we divide a square into four sections, called **W, X, Y,** and **Z**. At B, we reassemble the four sections into a rectangle with an area that appears 1 square unit larger than the area of the original square.

PROBLEM 5-5

Some flaw must exist in the rearrangement process shown in Fig. 5-4. Where does the "phantom square unit" come from?

SOLUTION

The trouble exists in the center of the rectangle of Fig. 5-4B. We have a small 1-by-3-unit rectangle there (Fig. 5-4C). We can slice this rectangle into two 1-by-3 right triangles, as shown. Immediately to the left and right of this central rectangle, we find two 2-by-5 right triangles. If the long diagonal through the 13-by-5 rectangle were a straight line, then the ratio 1:3 would equal the ratio 2:5. But these ratios differ, as we can easily verify with a calculator!

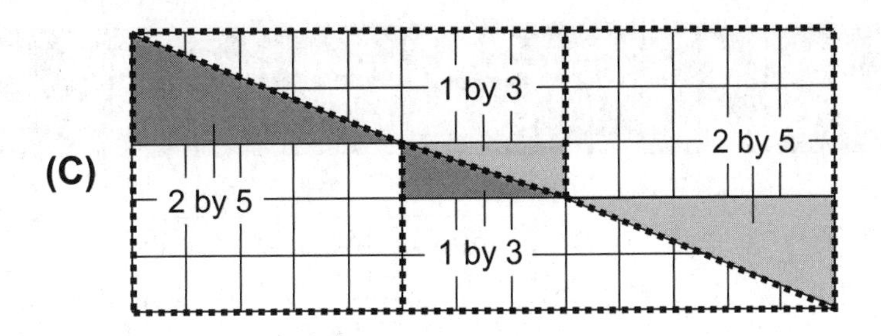

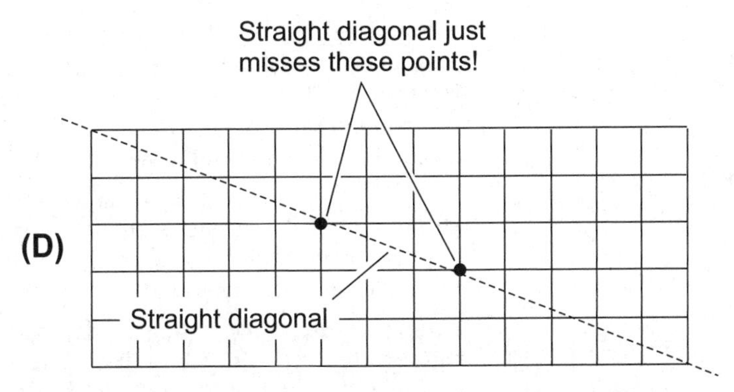

FIGURE 5-4 • (Continued) At C, illustration showing right triangles "constructed" within the rectangle. At D, an actual straight diagonal of the large rectangle doesn't intersect any of the vertices of the small squares.

The long diagonal cutting the 13-by-5 rectangle in Fig. 5-4C does not constitute a straight line segment. If it did, we would have the scenario of Fig. 5-4D, where the long diagonal line ramping down the 13-by-5 rectangle would not intersect the vertices (corners) of any of the small squares. The visual deception occurs over such an elongated shape that most people can't see it in an ordinary line drawing.

A "proof" that −1 = 1

We can "prove" that −1 = 1 in several different ways. All such "proofs" involve subtle reasoning flaws, or the neglect of certain facts of arithmetic. Table 5-1 shows one method of "proving" that −1 = 1. In this "proof," we denote the square root of a quantity as the quantity raised to the 1/2 power. That is, the 1/2 power means exactly the same thing as a radical sign for the purposes of this discussion. We start with an obviously true statement, and from it, apparently using the rules of arithmetic, we derive a falsehood.

TABLE 5-1 A "proof" that −1 = 1. The 1/2 power of a quantity denotes the square root of that quantity. Each line in the table proceeds from the previous line, based on the reason given.

Statement	Reason
$(-1)^{1/2} = (-1)^{1/2}$	A quantity is always equal to itself.
$[1/(-1)]^{1/2} = [(-1)/1]^{1/2}$	Both $1/(-1)$ and $(-1)/1$ are equal to −1. Either of these expressions can be substituted for −1 in the previous equation.
$1^{1/2}/(-1)^{1/2} = (-1)^{1/2}/1^{1/2}$	The square root of a quotient is equal to the square root of the numerator divided by the square root of the denominator.
$(1^{1/2})(1^{1/2}) = [(-1)^{1/2}][(-1)^{1/2}]$	Any pair of equal quotients can be cross-multiplied. The numerator of the one times the denominator of the other equals the denominator of the one times the numerator of the other.
$(1^{1/2})^2 = [(-1)^{1/2}]^2$	Either side of the previous equation consists of a quantity multiplied by itself. That is the same thing as the quantity squared.
$1 = -1$	When the square root of a number is squared, the result is the original number. Therefore, all the exponents can be taken out of the preceding equation.

PROBLEM 5-6

What's wrong with the "proof" portrayed in Table 5-1?

SOLUTION

Perhaps you wonder whether we have any business talking about the square root of a negative number. That's not the problem here! Mathematicians have defined the square roots of negative numbers, and the resulting quantities behave according to perfectly good rules of arithmetic (although they seem weird to people who've never worked with such quantities before). The problem with the "proof" shown in Table 5-1 lies in the fact that the square root of any quantity can be positive or negative.

We normally think that the square root of 1 simply equals 1. But the square root of 1 can also equal −1. If we multiply −1 by itself, we get 1. In a sense, then, the square root of 1 has two different numerical values "at the same time"! The trick in the "proof" shown by Table 5-1 exploits this

fact to deceive the unsuspecting nonmathematician. We "accidentally" suppose that both elements of the "two-valued" square root of 1 equal each other—in effect assuming the false conclusion of the "theorem"!

Wheel Paradox

Here's a famous paradox that concerns a pair of concentric wheels. Imagine that we attach two wheels, one large and one small, rigidly together with a common center so that one rotation of the large wheel always coincides with exactly one rotation of the smaller wheel. Suppose that we roll this "double-wheel" along a two-level surface, as shown Fig. 5-5, so that the large wheel makes exactly one rotation along the lower surface. If the diameter of the larger wheel equals d_1 units, then the length of the path along the lower surface for one rotation equals πd_1 units according to the rules of geometry.

The upper surface lies precisely the right distance above the lower surface, so the smaller wheel can move along the upper surface while the larger wheel moves along the lower surface. Suppose that the smaller wheel has a diameter of d_2 units. The smaller wheel rotates at the same rate as the larger wheel. When the larger wheel completes a full rotation, so does the smaller wheel. But the smaller wheel traverses the same distance (πd_1 units) as the larger wheel does, as we can see by examining Fig. 5-5. This coincidence implies that the two wheels have the same circumference, even though their diameters differ. That's impossible!

How can we resolve the foregoing paradox? Two wheels with different diameters cannot have the same circumference, can they?

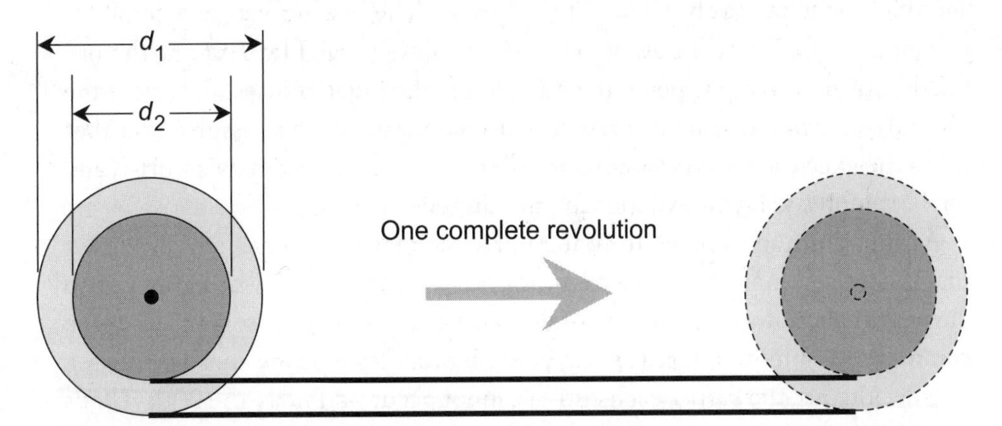

FIGURE 5-5 • The wheel paradox. The larger wheel has diameter d_1, and the smaller wheel has diameter d_2.

The catch lies in the fact that the smaller wheel must slide or skid along the upper surface when the larger wheel rolls along the lower surface with good traction. Or, conversely, the larger wheel must skid when the smaller wheel rolls with good traction. Nothing in the statement of the problem forbids wheel-skidding!

Classical Paradoxes

Ever since our distant ancestors began communicating with words, people have told each other stories in which fallacious logic could lead to improbable or incredible conclusions. The following few paradoxes have been passed down to us through the generations. Each one contains a logical flaw or fallacy. Can you identify the flaw in each of these classical paradoxes?

Execution Paradox

Imagine that you have been convicted of a horrible crime, and the court has condemned you to die as punishment. You live in a state where executions transpire without delay, and by a method that scientists have proven humane in the sense that the condemned person feels no physical pain. (We might wonder how anybody can claim to know whether or not a deceased person felt pain at the moment of death, but a discussion of that mystery belongs in another book.)

The judge sentences you on Friday the 13th. She tells you that your rendezvous with death will occur at noon on a weekday during the following week, and that you'll have only 10 minutes of notice. The executioner will come for you instead of the lunch boy, who normally shows up at 11:50 A.M. with your midday meal. Adding suspense to anticipation, the judge refuses to disclose the exact day on which your execution will take place, but she assures you that when the executioner arrives at your cell door, you'll get taken by surprise, and for 10 minutes you will experience pure disbelief.

As you return to your cell to contemplate your fate, you try to convince itself that the judge cannot seriously commit first-degree murder by ordering you to die with malice aforethought. No—she must know that your sentence is to be commuted to life in prison (or maybe something less, if luck runs your way). You decide that the surprise execution cannot occur on Friday the 20th. That's because, if you haven't been executed by noon on Thursday the 19th, you'll know that the date must be on the morrow; but then the event won't come as a surprise, and the judge told you that you'd get taken by surprise. That little bit of logic cuts the "execution week" a day short, encompassing only Monday the 16th through Thursday the 19th.

Continuing your ruminations in a sort of backward mathematical chain reaction, you conclude that your execution can't take place on Thursday the 19th because, if the executioner has not come for you by 11:50 A.M. on Wednesday the 18th, you'll know that the date must be on the morrow. But then, as in the previous step, the event will not constitute a surprise. Now you have cut the "execution week" down to only three days. You must die at noon on Monday the 16th, at noon on Tuesday the 17th, or at noon on Wednesday the 18th.

You repeat the above-described process twice more in the same fashion, eliminating Wednesday the 18th, and then Tuesday the 17th. That leaves only Monday the 16th as a possible date for your demise; but you quickly dispense with that day as well because, having "eliminated" every other day of the five-day work week, you won't feel any surprise when the executioner comes for you at 11:50 A.M. on Monday. Recalling that the judge said you'd get taken by surprise, you eliminate Monday the 16th and convince yourself that the judge gave you a coded message: Your sentence has been commuted to life in prison (or maybe something less).

Imagine your surprise—as the judge promised that you would experience—when, at 11:50 A.M. on Tuesday the 17th, the lunch boy fails to arrive with your midday meal, and instead the executioner knocks and motions for you to begin the 10-minute trek down the long, dark hallway to the execution chamber!

A Two-Pronged Defense

Most of us have had a checkout clerk, in a grocery store for example, test our currency for legitimacy. As cash becomes less common in favor of credit and debit cards (or other, more sophisticated technologies), experiences like this will doubtlessly become rare. But for now, the familiar sight remains: a service person marking a twenty-dollar bill with a special pen to see if the ink turns dark, which it almost never does.

Almost never.

But suppose that one day, as you listen to the "beep, beep, beep" of the laser scanner in the grocery store and watch package after package of your next week's food go by, the clerk stops, holds up a bill with a black mark across its face, and says, "This bill is counterfeit. In fact, all of the bills you've given me are fake." You live in a state where security and law-enforcement departments function with ruthless efficiency. Before you can utter a word in response, you find yourself shackled in handcuffs, hustled out of the store by two burly men in uniform while children stare as if you've just arrived from an alien planet. The charge: attempting to pass counterfeit currency.

You come in front of the judge to defend yourself. "I didn't know that all those bills were fake," you say. "And anyhow, I stole them. Some idiot made a big deal out of counting his cash on a street corner, and I swiped the bills right out of his hand and ran off. I would never have stolen bills that I knew, or even suspected, were counterfeit!" The judge takes into account the fact that you've never been arrested for anything before in your life; you haven't even had a traffic violation. The judge decides to drop the charge of attempting to pass counterfeit bills, but in its place, she imposes a new charge: Petty larceny. You did, after all, steal the bills from that idiot on the street corner.

You have a ready-made defense for this charge as well. You retort, "The bills I stole are all counterfeit. They're completely worthless. How can I be guilty of stealing something that has no value? In fact, you might say that I did that poor fool on the street a favor. Now he can't get arrested for trying to pass those counterfeit bills, because he doesn't have them anymore!" The judge looks at you with an expression that reminds you of the way the children in the store gawked as the cops led you out. She calls the prosecuting attorney to the bench, and together they whisper below the threshold of your hearing. Then the judge proclaims, "Case dismissed."

Do you think this scheme will work in the state, country, or planetary federation where you live? I have some advice for you: *Don't try it!* Few judges or prosecutors are as fallible, or as benevolent, as those in this little parable. And anyhow, with all the nefarious shenanigans that occur in today's cyber-punk paradise, you should count your blessings if you can stay out of jail even if you never commit so much as a traffic violation for the rest of your life.

Saloon Paradox

Imagine that you and I sit in a saloon full of partygoers. It's 11:30 on a Saturday evening, and every chair and stool has an occupant. Although we don't know how many people are drinking and how many are not, we can say, based on our new-found knowledge of predicate logic, that one or the other of the following statements must hold true right now:

- Everyone in the saloon is drinking.
- Someone in the saloon is not drinking.

First, let's assume the truth of the first statement: Everyone in the saloon is drinking. Then if any particular person is drinking, we can logically conclude that everyone is drinking (because, after all, we've already assumed that everyone is drinking!).

Now let's hypothesize the truth of the second statement: Someone in the saloon is not drinking. Using simple logic, we can conclude that if that particular "someone" is drinking, then everyone in the saloon is drinking. We've generated an artificial contradiction (a particular person is both drinking and not drinking), from which anything follows, including the conclusion that everyone in the saloon is drinking.

In either the first or the second instance, we can find someone in the saloon such that, if he or she is drinking, then everyone in the saloon is drinking. That's obviously ridiculous, because it suggests that we can never find a saloon in which some of the people are drinking and some of the people aren't.

Barbershop Paradox

Aloysius and Boris take a trip to the barbershop. Three barbers live and work in the shop: Mr. Xerxes, Mr. Yeager, and Mr. Zimmer. The three barbers share a common trait: They hate crowds. Because they adhere to the maxim "Three's a crowd," they never occupy their shop simultaneously.

Boris wants to get a trim from Mr. Zimmer. Boris knows that the shop is open, so at least one of the barbers must be available. Everyone in town knows that whenever Xerxes leaves the barbershop, he takes Yeager along with him.

Boris says, "I hope Mr. Zimmer is in the shop now."

"He is," says Aloysius.

"How can you possibly know that?" asks Boris.

"I can logically prove it," says Aloysius.

Boris says, "Okay."

Aloysius begins, "For the sake of argument, let's suppose that Zimmer is not in the shop. In that case, if Xerxes is also out, then Yeager is in, because at least one of those three guys must be there for the shop to be open."

Boris says, "Okay."

Aloysius continues, "We know that whenever Xerxes leaves the shop, Yeager leaves along with him. Therefore, if Xerxes is out, Yeager is out."

Boris says, "Okay."

Aloysius goes on, "If Zimmer is out, then we know that the statements 'If Xerxes is out then Yeager is in' and 'If Xerxes is out then Yeager is out' are both true.

Boris says, "Okay."

Aloysius says, "But those two statements contradict each other."

Boris says, "Okay."

Aloysius says, "We get a logical absurdity if we assume that Zimmer is out. Therefore, we know that Zimmer must be in the shop."

Boris says, "Okay."

Aloysius says, "Therefore, whenever the shop is open, Zimmer is there."

Boris says, "Zimmer's an okay guy, but I can't believe he works that hard."

Aloysius says, "According to the laws of mathematical logic, he must!"

"Somehow, that's not okay," says Boris. "How can the laws of mathematical logic govern the behavior of my favorite barber? Zimmer's one of the most illogical people I know."

Shark Paradox

You take a long swim in the ocean with your best friend, despite the warnings from the local folks that the waters harbor sharks with two unsavory habits: The big fish talk to people, and the big fish eat people.

As you're enjoying the cool surf on that hot afternoon, your friend suddenly goes under the water. Almost before you know what has happened, you hear a rasping voice: "We sharks have taken your friend down below, where we are going to eat her." You turn to see who would say such a thing and discover not a human, but a shark.

Thinking rationally, which you always do when confronted by eloquent sharks, you say, "Oh, you don't want her."

"Hmm," says the shark. "You might be right. I'll give her back to you on one condition. You must guess whether I'll live up to my word or not."

"What do you mean?" you ask.

"I promise to return your friend if and only if you can correctly predict whether or not I will keep that promise."

Now your brain begins to function in a more rational way than you've ever known it to operate. Not only have your logical powers peaked, but your sense of humor has spiked as well. You decide to trick the humanivorous creature. You say, "I predict that you will not return my friend."

Now what can the shark do? If he returns your friend, then your prediction turns out false, in which case the shark cannot return her. If the shark does not return your friend, then your prediction turns out true, in which case the shark must return her. Therefore, the shark will return your friend if and only if he does not.

"You are a strange human," says the shark. Just then your friend surfaces, unharmed. "Get away from us sharks," continues the big fish, "and don't ever come around this ocean again."

Russell and Gödel

Let's revisit the paradox involving Hap. If we assume that Hap exists, then he shaves himself and doesn't shave himself. An English logician, *Bertrand Russell* (1872–1970) expanded on this conundrum to obtain *Russell's paradox*. Along similar lines, a Czech-born mathematician named *Kurt Gödel* discovered in 1931 that in the entirety of mathematics, some statements remain *undecidable*. That means we can never prove them either true or false. We call this result the *Incompleteness Theorem*.

Sets

We'll learn the basics of set theory in Chap. 8. For now, we can define a *set* as a collection of things (such as numbers, physical objects, concepts, people, or even other sets) considered as a group. Some texts use the term *class* instead of set.

When we think about a collection of objects as a set, we give the entire group an identity of its own; the group itself becomes an object apart from the objects that make it up. For example, we might talk about all the trees in Wyoming in a general sense, in which case we create an idea; but if we consider the *set* of all trees in Wyoming, we invent a specific, single object. We might see a dozen eggs and think nothing special of them (other than the fact that eating them all at once might prove nauseating), but if we think of the collection as a set—in a carton on a shelf at your local grocery store, for example—the bunch of eggs takes on a new identity, quite apart from the eggs themselves.

Certain eggs can compose the *elements*, also called *members*, of a specific set, just as certain trees can compose the elements of another specific set. We could even consider the set comprising the two sets just described; that set would exist as an object with its own unique identity (although we might find ourselves hard-pressed to imagine any use for it).

Two Special Sets

Russell posed the question: Does any set exist, such that it forms an element of itself? Apparently, the answer is yes. Consider, for example, the set S_U of all sets in the universe—that is, the set of all sets that can possibly exist. We know that S_U is a set, and by default it must belong to the set of all sets (i.e., itself). The set of all sets with more than 10 elements is also an element of itself. If we think about this business for awhile, we can imagine infinitely many such sets.

Once Russell had ascertained the existence of sets that constitute members of themselves, he asked the opposite question: Does any set exist, such that it *is not* an element of itself? We have no trouble answering this question in the affirmative. The set of all trees in Wyoming is not itself a tree in Wyoming; the set of a dozen eggs in a carton is not itself an egg in a carton. Again, a little thought leads us to realize that infinitely many sets of this sort must exist.

Now imagine the set of *all sets in the universe* that are not members of themselves. Let's call this set S_N. We can reasonably suppose that such a set S_N exists. We know that S_N has at least two elements (the set of all trees in Wyoming, and the set of a dozen eggs in a certain carton in some grocery store near you). However, S_N does not encompass the set of all sets, because some sets exist that belong to S_U, and none of them can belong to S_N. Even if we cannot specifically name all the elements of S_N, or even define them all by implication, this inability doesn't preclude the existence of S_N.

The Paradox

Now that we have established a reasonable basis for believing that a set S_N actually exists, let's ask the question: Does S_N belong to itself? That is, can we say that S_N is an element of itself? This question goes deeper than we might at first imagine. It leads us straightaway into a contradiction from which even the keenest thinkers in history have failed to make a clean escape.

Let's assume that S_N is a member of itself. In that case, it meets the condition for belonging to S_N, namely, that it is not a member of itself. That conclusion contradicts our assumption, so *reductio ad absurdum* (RA) tells us that our assumption is false. Therefore, S_N is not a member of itself. But remember: We have defined S_N as the set of all sets that are not members of themselves. That means, in fact, S_N is an element of S_N, so it's an element of itself, which brings us back to our original assumption.

This paradox differs fundamentally from the situation with Hap in the town of Happyton. We can simply deny the existence of any barber such as Hap or any town such as Happyton. We cannot so easily deny the existence of the set of all sets that are not members of themselves.

Professor N's Machine

Imagine that the venerable Professor N, one of your favorite mentors from your radical college days, comes to you with a contraption that looks like a personal computer, but quite a lot larger and heavier. "This," says Prof. N, "is my universal truth determination machine."

"What's that?" you ask.

"A machine that can prove or disprove any statement you make," boasts Dr. N.

"Okay," you say. "What sort of program, what software, does it use?"

"I can't tell you that," says Dr. N, "because it's a trade secret. But I assure you that this computer's program has finite length."

"Well," you say, "I would think so, because the machine has finite size."

"Good observation," says Dr. N. "Now I dare you to say something that my machine can't prove or disprove."

You think for a moment, perhaps reflecting on Hap and Russell, and then say to the machine, "You'll never say that this sentence is true."

"What?" asks Prof. N.

"I just asked the machine to prove," you say, "that it will never say that this sentence is true."

The machine remains silent.

"I don't understand," says Prof. N.

You declare to the machine, "I have made a proposition to you. I have said that you, the wizard of logic, will never claim the truth of a certain statement. What is that proposition? Please simply restate it, and give it a name."

The machine says, "Your proposition is, 'I, meaning myself, the universal truth determination machine, will never say that this sentence is true.'" Then it adds, "I propose that we call this proposition N, after my creator, Dr. N."

"Okay," you say, and then you ask the machine, "Is N true or not?"

The machine says nothing.

"Let me help you," you say. "Oh wizard of logic, you must not say that N is true, because if you do, then you have asserted, 'I will never say that this sentence is true,' implying precisely the opposite—if you really are the wizard that your creator says you are."

Dr. N looks at you as if you have just arrived on a spacecraft from an alien planet. The machine remains quiet.

You continue, "Oh great and powerful algorithm processor, you must not say that N is false. If you do that, then you've actually made the statement N itself, "I'll never say that this statement is true," implying the existence of a statement about which you will never make a truth or falsity claim."

The machine remains silent.

Dr. N looks at the floor.

"My old friend," you say as you turn to the professor, "your machine cannot prove my proposition, and your machine cannot disprove my proposition."

What Did You Prove?

The foregoing hypothetical conversation outlines, in a simplified (high-level mathematicians would doubtless say oversimplified) way, the proof of Kurt Gödel's Incompleteness Theorem. You found a statement whose truth or falsity defies determination. Gödel's theorem actually claims that, within the framework of mathematics, undecidable statements such as N exist.

TIP ***Some lay people have taken Gödel's Incompleteness Theorem out of context to come up with fantastic "conclusions" having nothing to do with the scope or intent of Gödel's work (or that of any serious logician). For example, some people have suggested that the Incompleteness Theorem allows that the claim "God exists!" is undecidable, or that the statement "Extraterrestrial humanoids do not exist!" is undecidable. Gödel never intended to venture into such rarefied territories, contenting himself to deal in pure logic and mathematics—which, we now realize, is rarefied enough.***

QUIZ

You may refer to the text in this chapter while taking this quiz. A good score is at least 8 correct. Answers are in the back of the book.

1. **Suppose that we've successfully proved that any endless decimal number of the form**

   0.*nnnnnnn* ...

   constitutes a rational number, where *n* represents a single-digit whole from 0 to 9 inclusive. Based on that theorem, we get the idea that all endless decimal numbers must constitute rational numbers. We have committed

   A. the probability fallacy (PF).
   B. *reductio ad absurdum* (RA).
   C. the fallacy of "proof" by example.
   D. no fallacy.

2. **Someone tells you that they're 90% certain that Julius Caesar died as the result of wounds caused by a metal knife in the year 44 B.C. This statement represents an example of**

   A. the probability fallacy (PF).
   B. *reductio ad absurdum* (RA).
   C. the fallacy of "proof" by example.
   D. no fallacy.

3. **Suppose that you flip a coin six times, and it lands "heads up" five times. You conclude, based on that experiment, that the coin is "weighted" or "biased" in some physical way that causes it to land "heads up" more often than "tails up." This conclusion represents an example of**

   A. hasty generalization.
   B. the probability fallacy (PF).
   C. begging the question.
   D. fuzzy logic.

4. **What's the principal difference between Russell's paradox and the dilemma concerning Hap, the barber in the fictional town of Happyton?**

   A. Russell's paradox is more difficult to resolve.
   B. Hap's paradox is more difficult to resolve.
   C. We can resolve Russell's paradox by *reductio ad absurdum*, while Hap's paradox merely constitutes a hasty generalization.
   D. There's no difference. The two paradoxes are qualitatively the same.

5. **Which of the following sets is a member of itself?**

   A. The set of all the eggs in all the cartons in all the grocery stores in the world.
   B. The set of all sets that can exist.
   C. The set of all sets that cannot exist.
   D. None of the above, because no such set exists.

6. **How does inductive reasoning compare to deductive logic?**
   A. They're identical; the terms refer to precisely the same rigorous technique for proving theorems.
   B. Deductive logic constitutes an accepted rigorous method of proof, while inductive reasoning does not.
   C. Inductive reasoning constitutes an accepted rigorous method of proof, while deductive logic not.
   D. Both terms refer to logic games involving the probability fallacy (PF), because neither of them can offer absolute proof of anything.

7. **How can we explain the flaw in "proof"? that $-1 = 1$ shown in Table 5-1?**
   A. We can't, because it represents a genuine paradox.
   B. We can't, because it represents an example of an unprovable truth.
   C. It subtly overlooks the full meaning of a square root.
   D. It makes the notion of a square root unnecessarily complicated.

8. **Which of the following statements holds true concerning sets?**
   A. A set is precisely the same thing as the elements that make it up.
   B. A set can never be an element of itself.
   C. A set can never fail to be an element of itself.
   D. A set possesses an identity apart from its elements.

9. **Truth values can range along a smooth continuum in**
   A. hasty generalization.
   B. *reductio ad absurdum* (RA).
   C. begging the question.
   D. fuzzy logic.

10. **The demonstration of Russell's paradox involves an application of**
    A. hasty generalization.
    B. the probability fallacy (PF).
    C. *reductio ad absurdum* (RA).
    D. inductive reasoning.

chapter 6

Strategies for Proofs

Let's revisit the elements that make up a *mathematical theory*, also called a *mathematical system*. Then we'll look at some common techniques that theoreticians use to prove statements true or false in the context of the system. You got a taste of this methodology in Chap. 4. Now we'll expand on that theme.

CHAPTER OBJECTIVES

In this chapter, you will

- Envision the structure of a theory.
- Formulate definitions and axioms.
- Outline the fundamental principles of Euclidean geometry.
- Prove theorems in Euclidean geometry.
- Learn to use mathematical induction.
- Compare correlation, causation, and implication.

How Does a Theory Evolve?

We can build a mathematical theory from "self-evident facts" and formal definitions, applying the rules of logic to prove or disprove statements known as *propositions*. Once we've proven a proposition, we can use it to prove or disprove more propositions. As we continue this process, we accumulate an expanding set of truths. Hopefully, we'll never encounter a contradiction, and our set of truths will evolve into an interesting and elegant "thought universe."

Definitions

Imagine a mathematical theory as a large, complicated building. The definitions constitute the stones of the foundation. Without a solid foundation, we can't expect our structure to function; nobody will want to "live in it." Without adequate definitions, a mathematical theory will not survive the scrutiny of peer review, the challenge of applicability to the real world, and all the other assaults that theories must face. However, with sound and sufficient definitions, flawless logic, and a fair measure of good luck, a theory can evolve into something interesting and elegant, and it might make a major contribution to the body of human knowledge. Here are some examples of rigorous definitions:

- A *set* is a collection or group of objects called *elements*. A set can be denoted by listing its elements in any order, and enclosing the list in curly brackets (also called *braces*).
- The *empty set*, also called the *null set*, is the set containing no elements, symbolized { } or ∅. A set is empty if and only if it contains no elements.
- Let Q be a point in a flat, two-dimensional plane X. Let C be the set of all points in X that exist at a fixed distance *d* from Q, where *r* is not equal to 0. Then C is the *circle of radius d in X, centered at point* Q.
- Let Q be a point in a three-dimensional space Z. Let S be the set of all points in Z that exist at a fixed distance *d* from Q, where *d* is not equal to 0. Then S is the *sphere of radius d in Z, centered at* Q.

We can sometimes use drawings to help us portray definitions, but a truly rigorous definition must "carry its weight" without any illustrations. Figure 6-1A shows an example of a circle in a flat plane as defined above. Figure 6-1B shows an example of a sphere as defined above. While these drawings might help readers understand the respective definitions, we should not *have to* use the drawings to make the definitions complete.

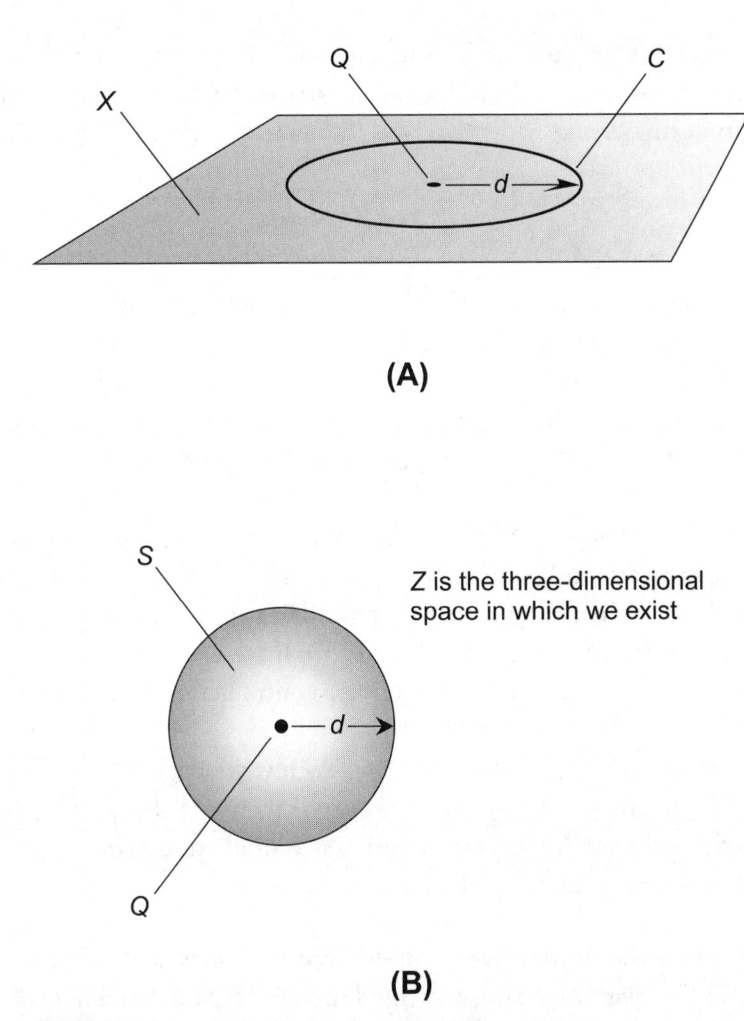

FIGURE 6-1 • At A, a circle in a plane. At B, a sphere in space.

Elementary Terms

A special sort of term, worthy of mention because it's as important as any of our standard defined terms, is the so-called *undefined term* or *elementary term*. Most mathematical theories have some of these. For example, in geometry the terms *point*, *line*, and *plane* are elementary. We can describe these objects so that readers can envision how they look in the real world, but it's difficult or impossible to *rigorously* define them. In set theory, we don't rigorously define the notion of a *collection* or *group* when talking about sets. When using elementary terms, we must assume that our listeners or readers have a reasonable amount of intelligence and insight. When we create a mathematical system, we should always strive to minimize the number of elementary terms.

Axioms

We can't derive any meaningful results without at least a couple of things that we accept without question. In a mathematical theory, we call these assumptions *axioms* or *postulates*. If we compare our definitions to the individual stones in the foundation of a building, then we can compare the set of axioms, taken all together, to the entire foundation, complete with mortar to hold the stones in place.

When we formulate a set of axioms to get a theory started, we should create as few axioms as we can get away with, while still allowing for a theory that produces plenty of meaningful truths. If we don't formulate enough axioms, then we can't expect to get much of a system. If we create too many axioms, we run the risk that we'll eventually derive a contradiction and see our theory collapse.

Once we have plenty of defined terms and a few good axioms, we can apply the rules of logic to generate a larger group of truths. If the set of axioms is *logically consistent*, then we will never derive a contradiction in our theory, even if we have a million years to work on it.

TIP ***When we decide on the axioms and write the definitions for a mathematical theory, we can minimize the risk of contradiction by keeping the number of axioms to the minimum necessary so that the theory makes sense and produces enough provable truths to be interesting.***

Euclid's Postulates

Some classical examples of axioms follow. You might recognize them as modified versions of the postulates set forth by the geometer *Euclid* in the third century B.C. We won't try to rigorously define *point*, *line segment*, *line*, *circle*, *center*, *radius*, *right angle*, *interior angle*, and *straight angle*. They're elementary terms. You know them from basic geometry anyway, don't you?

- Any two points P and Q can be connected by a line segment (Fig. 6-2A).

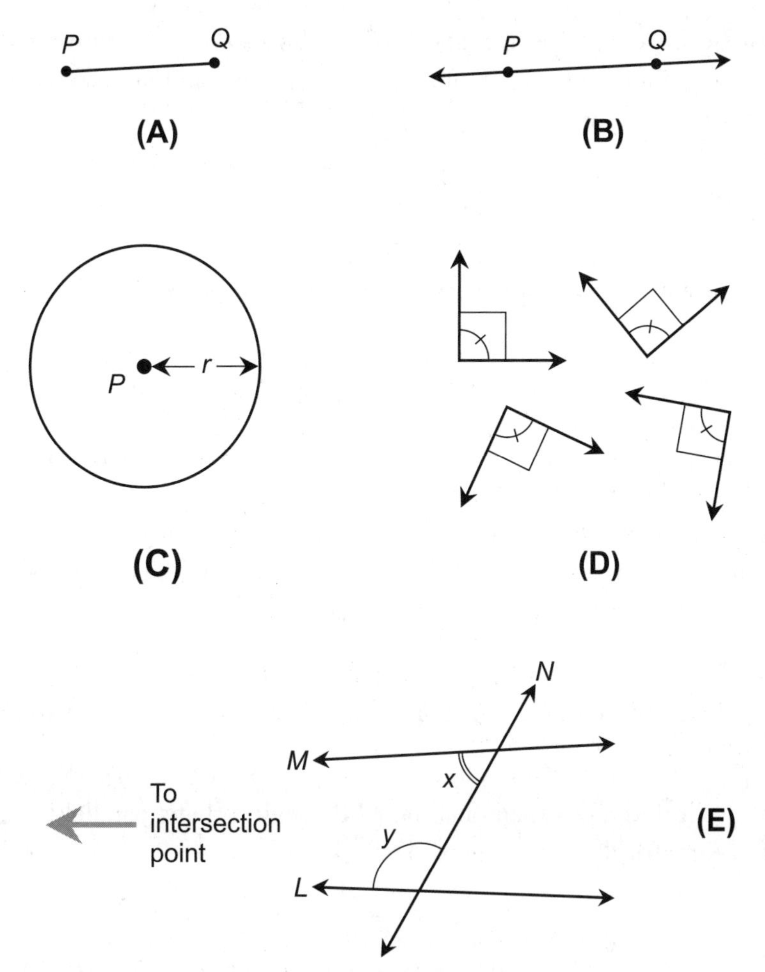

FIGURE 6-2 • Concepts behind Euclid's axioms in classical geometry.

- Any line segment can be extended indefinitely in both directions to form a line (Fig. 6-2B).
- Given any point *P*, we can construct a circle with *P* as its center and having radius *r* (Fig. 6-2C).
- All right angles are identical (Fig. 6-2D).
- Suppose that two lines *L* and *M* lie in the same plane. Also suppose that both lines are crossed by the same line *N*. Further suppose that the measures of the two adjacent interior angles *x* and *y* sum up to less than a straight angle (180°). Then lines *L* and *M* intersect on the same side of line *N* as we define the angles *x* and *y* (Fig. 6-2E).

Mathematicians call these five axioms *Euclid's postulates*. As with definitions, good axioms should not rely on illustrations, but you can make sketches if you find them helpful in understanding the concepts.

A Denial

The last of the above-stated axioms, *Euclid's fifth postulate*, is logically equivalent to the following principle called the *parallel postulate*:

- Let L be a line, and let P be some point not on L. Then there exists exactly one line M passing through P and parallel to L.

The parallel postulate has received enormous attention by mathematicians in the last few hundred years, especially *Carl Friedrich Gauss* (1777–1855) and *Bernhard Riemann* (1826–1866). If we deny the truth of this axiom, we get consistent mathematical systems that work just as well as the more familiar system of *Euclidean geometry*. Here's the denial:

- Let L be a line, and let P be some point not on L. Then the number of lines M passing through P and parallel to L can equal any nonnegative integer.

This *modified parallel postulate* allows for the possibility that, in some geometric systems, we'll find more than one line M through P that's parallel to L, while in other geometrical systems we won't find any.

Still Struggling

At first thought, many people have trouble comprehending the modified parallel postulate. But imagine what happens when you try to draw parallel lines on the surface of a sphere! If we define the "straight" path between any two points as the shortest possible distance between those points, then on a sphere, a "geometric line" takes the form of a circle with its center at the sphere's center. Any two such circles intersect at *antipodes* (a pair of points opposite each other) on the sphere. Gauss and Riemann might have feared that contradictions would result from their bold notions. But they went ahead with their theories, and such misfortunes never befell them. Nowadays, students can take courses in *Gaussian geometry* and *Riemannian geometry* at colleges and universities worldwide.

Propositions

Once we've written up a good set of definitions and axioms, we're ready to start building a mathematical theory. Do you want to devise a new type of number system? Do you want to invent a new way to think of sets? Do you have an idea that you'd like to pursue, such as the notion of numbers that can have more than one value "at the same time"? No matter what the objective, and no matter what the context, the process of theory-building involves using elementary terms, definitions, and axioms to prove propositions according to the rules of logic.

Theorems

Once we've proven a proposition, we call it a *theorem*. As the number of valid theorems increases, a mathematical theory becomes "richer," provided that no two theorems contradict. (We need not worry if a theorem holds true in one mathematical system but not in another, entirely different system.) If we can prove a statement P along with its denial ¬P within the same mathematical theory, then that theory reveals itself as *logically unsound* or *logically inconsistent*—a fatal flaw!

PROBLEM 6-1

Define and denote the term *interior angle* as it pertains to a triangle.

SOLUTION

Consider three distinct points *P*, *Q*, and *R*. Suppose that these three points do not all lie on a single line. Consider the following three straight line segments:

- **Line segment *PQ* with end points *P* and *Q***
- **Line segment *QR* with end points *Q* and *R***
- **Line segment *PR* with end points *P* and *R***

These three line segments form a triangle that we denote as ΔPQR. We define the three *interior angles* of ΔPQR as follows:

- **The smaller of the two angles at point *P*, where *PQ* and *PR* meet; we can denote this angle as $\angle RPQ$ or $\angle QPR$**

- **The smaller of the two angles at point *Q*, where *PQ* and *QR* meet; we can denote this angle as $\angle PQR$ or $\angle RQP$**
- **The smaller of the two angles at point *R*, where *PR* and *QR* meet; we can denote this angle as $\angle QRP$ or $\angle PRQ$**

A Classical Theorem

Here's an example of a theorem that most people know by the time they graduate from high school:

- Imagine three distinct points *A*, *B*, and C. Let ΔABC be the triangle with points *A*, *B*, and C as its vertices. Suppose that the sides of ΔABC are all straight line segments, and they all lie in the Euclidean (flat) geometric plane defined by points *A*, *B*, and C. Let *a*, *b*, and *c* be the lengths of the sides of ΔABC opposite points *A*, *B*, and C, respectively. Suppose that the interior angle at vertex C, symbolized $\angle ACB$, is a right angle. Then *a*, *b*, and *c* relate to each other as follows:

$$a^2 + b^2 = c^2$$

Figure 6-3 illustrates a triangle of this sort. We call it a *right triangle* because one of the interior angles is a right angle. The above equation has become known

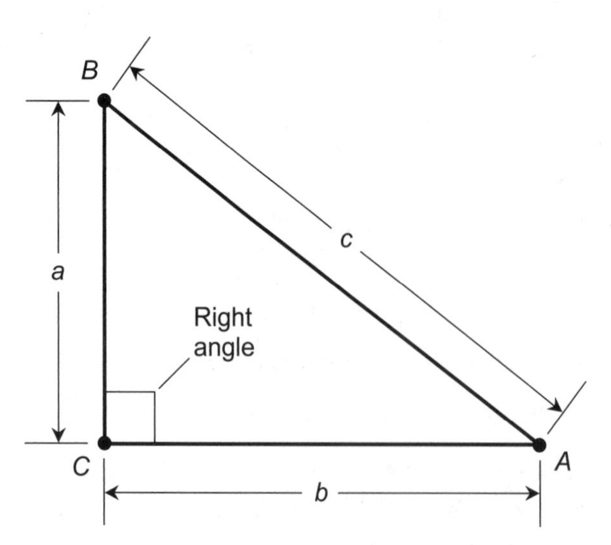

FIGURE 6-3 • The Pythagorean theorem tells us the relationship among the lengths of the sides of a right triangle that lies in a Euclidean plane. In this case, $a^2 + b^2 = c^2$.

as the *Theorem of Pythagoras*, named after a fifth-century-B.C. mathematician who, according to popular legend (which some historians doubt), rigorously proved it before anybody else did. Some people call it the *Pythagorean theorem*.

TIP ***We can derive this theorem from the fundamental axioms and definitions of Euclidean geometry, but it holds true only on flat surfaces. The Pythagorean equation does not work for "triangles" that lie on curved surfaces such as spheres, cones, ellipsoids, or hyperboloids.***

Lemmas

When we find ourselves confronted with a long, complicated, or difficult proof, we can streamline the logical process if we can prove one or more "preliminary theorems" called *lemmas* beforehand. We use the lemmas to help us prove the "final theorem." For this trick to work, we must decide in advance what the lemmas should tell us.

Once we've proven a lemma, we can save it for possible reuse in proving theorems to come. We should never discard the proof of any lemma, no matter how obscure it might seem at the moment. Years, decades, or centuries later, mathematicians might find that lemma valuable.

Corollaries

Once in awhile, after we've proven a theorem, a few short logical steps can lead us directly into the proof of another theorem. We call this sort of "secondary theorem" a *corollary*.

Consider the equation for the Pythagorean theorem. It holds true as long as we stay in a single Euclidean (flat) plane, but it doesn't always hold true on a non-Euclidean surface. We can state the following proposition and claim it as a corollary to the Pythagorean theorem:

- Let A, B, and C be three distinct points on a surface S. Consider the triangle formed by these points, symbolized ΔABC. Suppose that ΔABC lies entirely on S. Let a, b, and c be the lengths of the sides of ΔABC opposite the points A, B, and C, respectively. Suppose that the interior angle at vertex C, symbolized $\angle ACB$, is a right angle. If $a^2 + b^2 \neq c^2$, then S is not a Euclidean plane.

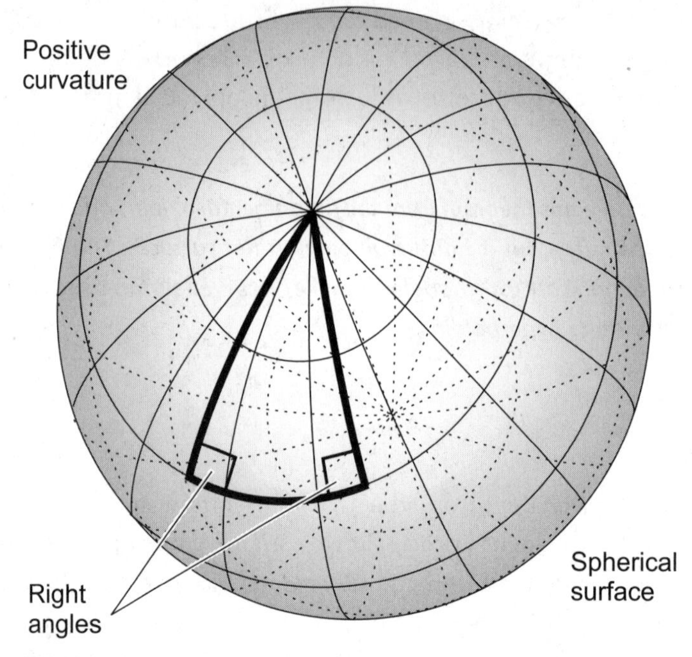

FIGURE 6-4 • A right triangle on a positively curved surface.

We prove this proposition using the law of the contrapositive. The statement "If *S* is a Euclidean plane, then the Pythagorean theorem equation holds true on *S*" is logically equivalent to the statement "If the Pythagorean theorem equation doesn't hold true on *S*, then *S* isn't a Euclidean plane."

Figures 6-4 and 6-5 show examples of right triangles on non-Euclidean surfaces. In Fig. 6-4, the surface has *positive curvature*; it bends in the same sense no matter what the orientation. Note that in this particular case, the triangle has two right angles inside! That state of affairs can never occur in a Euclidean plane. In Fig. 6-5, the surface exhibits *negative curvature*; it bends in one sense for some orientations and in the opposite sense for other orientations.

PROBLEM 6-2

Let's define a *conventional angle* as an angle that subtends an arc of less than half of a circle. Define and denote the term *measure of a conventional angle in degrees*.

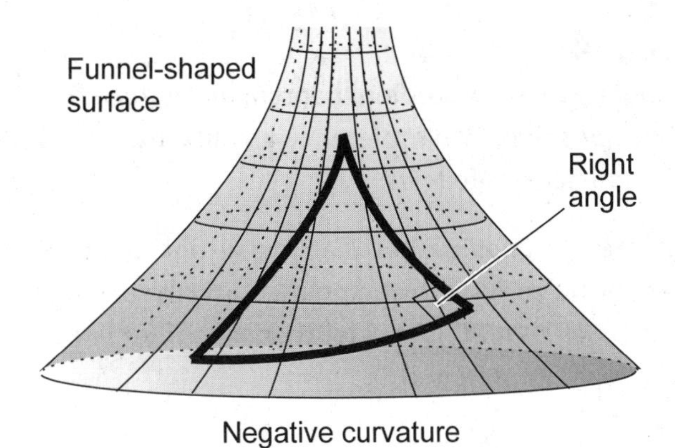

FIGURE 6-5 • A right triangle on a negatively curved surface.

SOLUTION

Consider three distinct points *P*, *Q*, and *R* and the following two straight line segments:

- **Line segment *PQ* with end points *P* and *Q***
- **Line segment *QR* with end points *Q* and *R***

Consider the smaller of the two angles at point *Q*, where *PQ* and *QR* meet. Call this angle $\angle PQR$. The *measure of the conventional angle $\angle PQR$ in degrees*, symbolized m°$\angle PQR$, is the fraction of a circle that $\angle PQR$ subtends, multiplied by 360.

PROBLEM 6-3

A familiar theorem in geometry states that the interior angles of a triangle always add up to 180° if the triangle and its sides lie entirely on a Euclidean (flat) surface. State this theorem formally without illustrations. Here's a hint: To shorten the statement, refer to the situation described in the solution to Problem 6-1.

SOLUTION

Consider three points, three line segments, a triangle, and its three interior angles as defined in the solution to Problem 6-1. If ΔPQR lies entirely on a Euclidean surface, then

$$m°\angle PQR + m°\angle QRP + m°\angle RPQ = 180°$$

PROBLEM 6-4

State a corollary to the preceding theorem that we can derive using the law of the contrapositive. Don't reference any illustrations. Let's introduce a new definition here, as follows:

- **On a surface *X*, a *geodesic* between two distinct points *P* and *Q* is the line segment or curve *PQ* with end points *P* and *Q*, such that *PQ* lies entirely on *X*, and such that *PQ* is shorter than any other line segment or curve on *X* with end points *P* and *Q*.**

SOLUTION

Consider three distinct points *P*, *Q*, and *R*, all of which lie on a surface *X*. Suppose that these three points do not all lie on a single geodesic. Consider the following curves:

- **Geodesic *PQ* with end points *P* and *Q***
- **Geodesic *QR* with end points *Q* and *R***
- **Geodesic *PR* with end points *P* and *R***

These geodesics form a triangle ΔPQR. Let's define the *interior angles* of ΔPQR as we did in the solution to Problem 6-1. Now imagine that one or the other of the following statements holds true:

$$m°\angle PQR + m°\angle QRP + m°\angle RPQ > 180°$$

or

$$m°\angle PQR + m°\angle QRP + m°\angle RPQ < 180°$$

In either of these situations, the surface *X* is non-Euclidean.

Proofs, Truth, and Beauty

Once we've demonstrated the truth of a few theorems in a mathematical system, we're likely to think of a lot of propositions that we'd like to prove. It's up to us, the mathematicians, to execute the logical steps in each proof with sufficient rigor to ensure the validity of the result.

Once in awhile, a proposition will appeal to the imagination, but we'll find ourselves unable to prove it. In a case like that, we face three possibilities:

- The proposition is false

- The proposition is true but *difficult* to prove
- The proposition is true but *impossible* to prove

A good mathematical theory can develop into an elaborate, fascinating, and (some would say) beautiful structure of theorems and corollaries, following a process such as that shown in the flowchart of Fig. 6-6. As long as no one finds a contradiction, the theory can grow indefinitely.

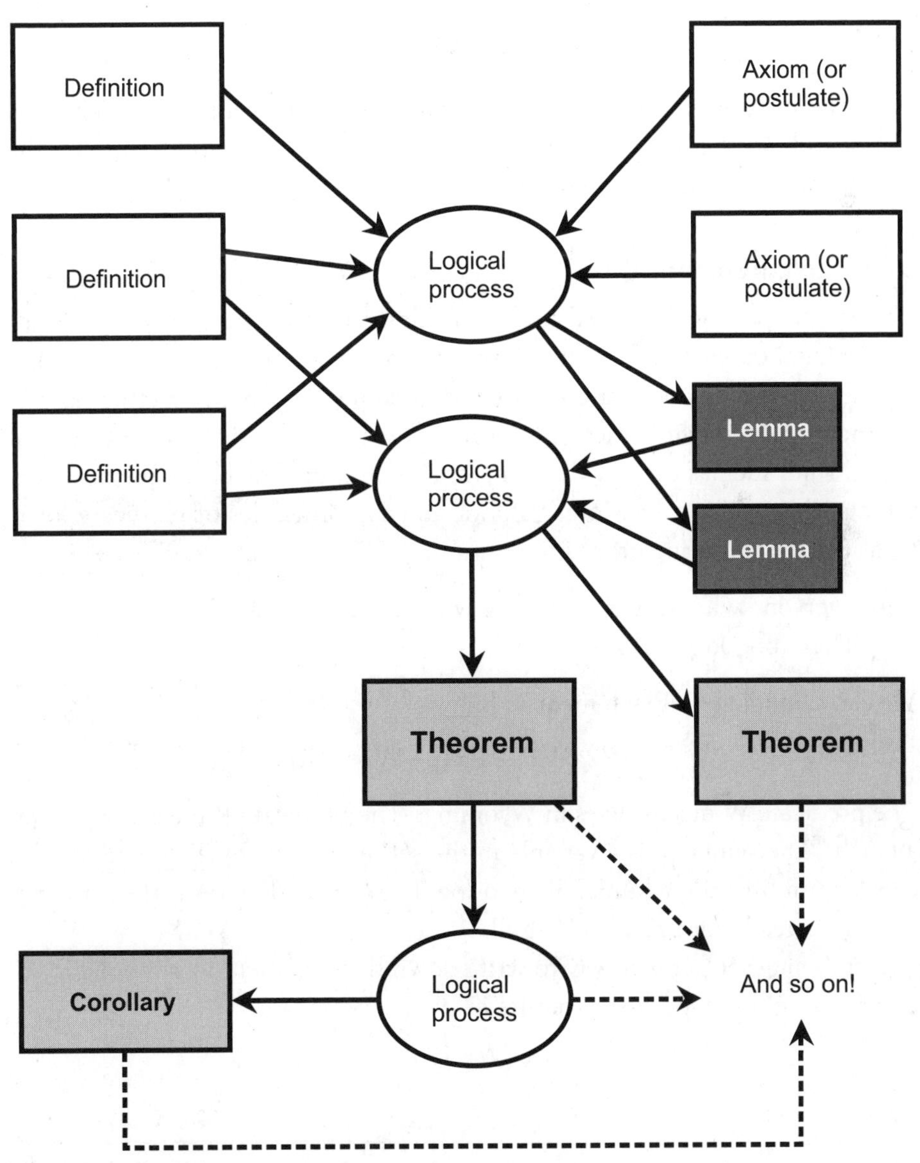

FIGURE 6-6 • Here's how a mathematical theory can evolve.

TIP ***According to the English mathematician G. H. Hardy (1877–1947), mathematical truths exist independently of the physical universe. If you're seriously interested in the philosophy behind the pursuit of pure mathematics, I recommend that you read Hardy's book*** **A Mathematician's Apology.** ***You should find it at any large public or university library. You can also order it from your local bookseller or buy it from an online source.***

Techniques

Over the centuries, mathematicians have developed an arsenal of powerful tactics to help them prove new theorems. All of these methods employ the rules of propositional and predicate logic.

Deductive Reasoning

Deductive reasoning, also called *deduction*, offers the most straightforward means of proving theorems. This scheme differs from the "process of elimination" that people sometimes use to argue in favor of something by discounting all the alternatives. In mathematics, the term *deduction* refers to the application of logical rules such as the law of the contrapositive, the law of double negation, the distributive laws, DeMorgan's laws, and the principles of predicate logic. Consider the following proposition:

- Everyone who lives in Wyoming works on a ranch. Joe lives in Wyoming. Therefore, Joe works on a ranch.

We can symbolize this statement as follows:

$$\{[\forall x\ (\mathrm{W}x \rightarrow \mathrm{R}x)] \wedge \mathrm{W}j\} \rightarrow \mathrm{R}j$$

The predicate W means "lives in Wyoming." The predicate R means "works on a ranch." The symbol x is a variable in the set of all humans. The symbol j is a constant, in this case a human named Joe. The upside-down A is the universal quantifier, read "For every" or "For all."

Here's another popular way to write down this argument in symbolic form, emphasizing the step-by-step nature of it:

$$\forall x\ (\mathrm{W}x \rightarrow \mathrm{R}x)$$

$$\mathrm{W}j$$

$$\therefore$$

$$\mathrm{R}j$$

In this set of statements, we list the progressive facts, one below the other. The three-dot symbol means "therefore." We can also write this formula as follows:

$$\forall x(Wx \to Rx)$$
$$Wj$$
$$\overline{}$$
$$Rj$$

Here, we've replaced the three dots with a horizontal bar, so the argument looks a little bit like a sum in arithmetic. You'll recognize this format from Chap. 3.

The foregoing proof illustrates an example of deductive reasoning: If a proposition holds true in general for a variable in a set, then that proposition holds true for every individual element of the set.

What's the Universe?

Before you try to prove propositions in a mathematical system, you ought to know the context or setting—that is, what you want to prove things about! Do you intend to prove a proposition involving the set of all the zebras in Africa? The set of all whales in the oceans? The set of all stars in the Milky Way galaxy?

We define the *universal set*, also called the *universe*, as the set of all objects to which we want a particular theorem or group of theorems to apply. Consider this statement: "If x is a real number, then $x + 1$ is also a real number." Here, our universe is the set of all real numbers. Now consider the following proposition: "The sum of the measures of the interior angles of any triangle on any Euclidean plane equals 180°." When we talk about the universe in this situation, we refer to the set of all triangles on all possible Euclidean planes.

Sometimes we can infer the universe from the context or setting, so we don't have to go to the trouble of specifying the universe. If we talk about people in general, the term "universe" refers to the set of all human beings; we know that fish or birds or caribou or squirrels aren't included. If we can't figure out the universe from context, then we must specify it. For example, we might preface a proposition with the following sentence: "Let x be a positive rational number." Or this: "Let T be a triangle on the surface of the beach ball in your swimming pool."

Weak Theorems

Suppose that you want to prove something for some, but not necessarily all, of the objects within a certain universe S. For example, consider the following proposition:

- Let W be the set of all widgets. Let D be the set of all doodads. There exists some element w in set W, such that w is also an element of set D.

In set theory, we can denote the phrase "is an element of" by writing $\in$. This symbol resembles the Greek letter epsilon, or a mutated English uppercase letter E. Given this symbol, we can write the above theorem as follows, based on the knowledge of what the letters stand for:

$$(\exists w)\ [(w \in W) \wedge (w \in D)]$$

From context, S refers to the union (i.e., the combination) of the set of all widgets and the set of all doodads. That is

$$S = W \cup D$$

where $\cup$ symbolizes set union. (You'll learn more about set theory in Chap. 8.) To prove a proposition of this form, we only need to provide one example for which the statement is true. Once we have shown that this type of proposition holds true for one object, we have shown that it holds true "for some" objects. That is, "there exists" an object for which the proposition is true.

We can call a theorem of this sort a *weak theorem*. A result like this might not seem profound when taken literally, but a gem of relevance may lie within, waiting for an eccentric genius to discover it some day!

Demonstrating a Weak Theorem

Consider this statement: "There exists a natural number that we can divide by 7, producing a quotient that equals another natural number." Suppose that we're interested in the natural number 765. Can we divide it by 7 to get another natural number? We can easily find out with a calculator. If your calculator agrees with mine, then you'll get

$$765/7 = 109.285714285714285714 \ldots$$

That quotient is obviously not a natural number. Now let's test the natural number 322. If your calculator agrees with mine, then you'll get

$$322/7 = 46$$

That quotient is a natural number. Now we know that there exists at least one natural number n such that, when we divide n by 7, we get another natural number. We also know that this property does *not* hold for *all* natural numbers.

When you want to demonstrate that a certain constant satisfies an existential proposition (i.e., one of the form "There exists ..." or "For some ..."), all you have to do is test the proposition and hope that the test comes out positive. If the test fails for a particular constant, you haven't disproved the proposition in general. In fact, you'll likely find it difficult or impossible to rigorously disprove an existential proposition.

Suppose that your friend tells you there's a 12-legged, bird-eating spider somewhere in South Dakota. If you can find one such animal, you'll have proven that your friend is correct. Of course, you'll be inclined to believe that your friend is wrong, because you've never heard of 12-legged bird-eating spiders living anywhere on earth, let alone in South Dakota. But how are you going to *prove* that there isn't at least one such brute?

Strong Theorems

In mathematics, we'll often encounter propositions that claim certain facts for all objects within a certain universe S. Once we've proved such a proposition, we can call it a *strong theorem*. For example, consider the following:

- Let R represent the set of all rational numbers. For all objects x, if x is an element of R, then $(x + 1)$ is an element of R.

Symbolically, we can write

$$(\forall x) \{(x \in R) \rightarrow [(x + 1) \in R]\}$$

Now consider the following proposition:

- Let T represent the set of all triangles on flat surfaces. Let H represent the set of all triangles for which the sum of the measures of the interior angles equals 180°. For all objects y, if y is an element of T, then y is an element of H.

Symbolically, we can write

$$(\forall y)\ [(y \in T) \rightarrow (y \in H)]$$

To prove a proposition of this form, we must rigorously demonstrate that for *any object* in the specified universe, the proposition holds true. We cannot get away with merely demonstrating the truth of the proposition for one object, or a few objects, or a thousand objects, or even a million objects. We must show that the proposition holds true for *all* of the objects in the universal set *without a single exception*.

Demonstrating a Strong Theorem

Consider the statement "All rational numbers are real numbers." Suppose we know that this statement is true. Let's think about the number −57/84. Is it a real number? If we can show that −57/84 is a rational number, then we can conclude that it's a real number. If we can't show that −57/84 is a rational number, however, our failure doesn't necessarily demonstrate that −57/84 is *not* a real number.

If you've taken first-year algebra, you know what a rational number is. Here's a formal definition:

- A number x is a *rational number* if and only if x can be expressed in the form a/b, where a is an integer, b is a natural number, and $b \neq 0$.

In case you've forgotten the formal definitions for *natural number* and *integer*, here they are:

- A number b constitutes a natural number if and only if b is an element of the set $N = \{0, 1, 2, 3, 4, ...\}$.
- A number a constitutes an integer if and only if a is a natural number or $-a$ is a natural number.

Let's consider *number* an elementary term for the purpose of this discussion. The following can serve as an informal definition of a real number:

- A number x is a *real number* if and only if $x = 0$, or x can express the distance between two points, or $-x$ can express the distance between two points.

Obviously, −57/84 is of the form a/b, where a is an integer, b is a natural number, and $b \neq 0$. We can simply let $a = -57$ and $b = 84$. The negative of −57 equals 57, and 57 is a natural number; therefore a is an integer. The number 84 is a natural number, because 84 is an element of the set $N = \{0, 1, 2, 3, 4, ...\}$. It's obvious that 84 is not equal to 0. Therefore, according to the original proposition, –57/84 constitutes a rational number. Because we've been told that all rational numbers are real numbers, we can conclude that the quantity –57/84 is a real number.

Let the predicate Q stand for "is a rational number." Let the predicate R stand for "is a real number." Let x be a logical variable. Let $k = -57/84$. Then we can write our single-instance proof like this:

$$\forall(x)\ \mathrm{Q}x \rightarrow \mathrm{R}x$$
$$\mathrm{Q}k$$
$$\overline{}$$
$$\mathrm{R}k$$

Reductio ad Absurdum Revisited

As you've learned, one of the most powerful tactics in the mathematician's theorem-proving arsenal is *reductio ad absurdum* (RA). In order to use this technique, we assume that the proposition we want to prove is false. Then we use the rules of logic to derive a contradiction from that assumption. That exercise proves that our original proposition is *not false*—so it must hold true!

Some purists argue that we should resort to RA only after we've tried and failed for a long time to prove a proposition by direct logical methods. But RA is a perfectly legitimate logical tool, and scenarios occasionally arise that seem to invite its use straightaway. In particular, statements of the form "There exist no ..." make ideal candidates for RA. If we find ourselves faced with the task of disproving an existential proposition—the so-called "proof of a negative"—then RA sometimes "cuts through the knot" with ease.

We've defined rational numbers, and we have a good idea of what constitutes a real number. Some real numbers are not rational. We call such numbers *irrational* and define them as follows:

- A number x is an *irrational number* if and only if x is a real number and x is not a rational number.

Now consider the following general proposition, which calls upon us to undertake a "proof of a negative":

- No irrational number can be expressed as an integer divided by a nonzero natural number.

In order to prove this proposition, let's assume that it's false, and let's call this new proposition A, as follows:

- A = There exists an irrational number that can be expressed as an integer divided by a nonzero natural number.

To use RA, we must try to derive a contradiction from A. Suppose that y is an irrational number that we can express as an integer divided by a nonzero natural number. Then $y = a/b$, where a is an integer and b is a nonzero natural number. It follows that y is rational, because y fulfills the definition of a rational number. But we just got done specifying that y is irrational (not rational)! We've used logic to derive a contradiction, so we have no choice but to conclude that A is false:

- It is not true that there exists an irrational number that can be expressed as an integer divided by a nonzero natural number.

This statement, ¬A, is logically identical to the original proposition, which we'd better restate to be sure that we haven't gotten disoriented during this convoluted process:

- No irrational number can be expressed as an integer divided by a nonzero natural number.

PROBLEM 6-5

Use RA to prove that there exists no largest rational number.

SOLUTION

Let's imagine that a largest rational number actually exists, and let's call it *r*. Our goal is to prove that this assumption leads to a contradiction.

It makes sense to suppose that *r* must have a positive value (as opposed to a negative or zero value), because it's larger than any other rational number! According to the definition of a rational number, we can break *r* down into a quotient *a*/*b*, where *a* represents an integer and *b* represents a nonzero natural number. Because *r* is positive, we know by definition that $r > 0$, so it follows that $a > 0$. (If *a* were negative, then *r* would have to be negative, and if *a* were equal to 0, then *r* would have to equal 0; we've ruled both of those possibilities out.) Now consider a number *s* such that

$$s = (a + 1)/b$$

We know that $a + 1$ is an integer, because 1 plus any integer always equals another integer. By definition, then, $(a + 1)/b$ is a rational number, and therefore s is a rational number. We also know that $(a + 1) > a$. Because $a > 0$, it follows that

$$(a + 1)/b > a/b$$

This inequality tells us that $s > a/b$, and therefore that $s > r$. Now we know that s is rational, and we also know that $s > r$. These two facts, taken together, contradict our assumption that r is the largest rational number.

We have no choice but to conclude that whatever rational number r we choose, no matter how huge, it's not the largest one. In other words, there exists no largest rational number.

Mathematical Induction

In some situations, we can prove propositions about all the elements in an infinite set, while carrying out only a finite number of steps. This scheme, known as *mathematical induction*, works with *denumerably infinite* sets. In this sort of infinite set, we can pair off the elements in a *one-to-one correspondence* with the natural numbers. We can always write the elements of a denumerably infinite set in the form of an "implied list."

Imagine a denumerably infinite set S consisting of elements s_0, s_1, s_2, s_3, s_4, and so on, such that we can portray the set as the "implied list"

$$S = \{s_0, s_1, s_2, s_3, s_4, \ldots\}$$

Suppose we want to prove that a certain proposition P holds true for all the elements of S. Imagine that we can prove both of the following statements:

- P holds true for s_0, the first element in S.
- If P holds true for some unspecified element s_n in S, then P also holds true for the next element $s_{(n+1)}$ in S.

By establishing the above two statements as facts, we spawn a "chain reaction of truths." We know P holds true for the first element in S, and this proves that P also holds true for the second element; that fact in turn proves P for the third element; the process continues without end. It's as if we've lined up infinitely many dominoes so that, when we knock over the first one, it falls against the second one and topples it; the second one falls against the third and topples it; the third topples the fourth; the fourth knocks down the fifth; the nth brings down the $(n + 1)$st—and we have a perpetual motion scenario! In the real world, the process might go on for thousands, millions, or billions of dominoes, lined up from Los Angeles to Las Vegas, from Seattle to St. Louis, from Vienna to Vladivostok. In the "mathematical world," the process can continue *forever*, going on to the edge of the known universe and beyond.

TIP ***Mathematical induction differs substantially from inductive reasoning, about which we learned in Chap. 5. Logicians accept mathematical induction as a valid method of deductive reasoning, powerful enough to offer "theorem-grade proofs" of propositions.***

PROBLEM 6-6

Show that for any two distinct rational numbers, there exists a third rational number whose value lies between them. Don't invoke RA, and don't try to use mathematical induction. You may use all the general rules of arithmetic (sums, products, differences, and quotients), however.

SOLUTION

Let's call the two rational numbers r and s. Suppose that the following equations both hold true:

$$r = a/b$$

and

$$s = c/d$$

where a and c represent integers, and b and d represent nonzero natural numbers. We know that such numbers a, b, c, and d exist, because r and s both constitute rational numbers, and the definition of a rational number requires that there exist such numbers a, b, c, and d.

Consider the *arithmetic mean* (more commonly called the *average*) of r and s. Let's call it x. We know that x lies between r and s, because the

arithmetic mean of any two numbers always ends up between them. According to the definition of the arithmetic mean, we have

$$x = (r + s)/2$$

When we substitute a/b for r and c/d for s, we obtain

$$x = (a/b + c/d)/2$$

From arithmetic, the general rule for a sum of two quotients allows us to rewrite the numerator of the fraction on the right-hand side of the above equation as

$$(ad + bc)/bd$$

Therefore, we know that

$$\begin{aligned}(a/b + c/d)/2 &= [(ad + bc)/bd]/2 \\ &= (ad + bc)/(2bd)\end{aligned}$$

and therefore that

$$x = (ad + bc)/(2bd)$$

If we can show that $ad + bc$ represents an integer and that $2bd$ represents a nonzero natural number, then we'll have proved that x must represent a rational number. Note the following arithmetic facts:

- The product of any two integers is always an integer. Therefore, ad and bc are integers.
- The sum of any two integers is always an integer. Therefore, $ad + bc$ is an integer.
- The product of any two nonzero natural numbers is a nonzero natural number. Therefore, bd is a nonzero natural number.
- Twice any nonzero natural number is a nonzero natural number. Therefore, $2bd$ is a nonzero natural number.

Taken all together, these facts tell us that x equals an integer divided by a nonzero natural number. By definition, x represents a rational number. As previously stated, x equals the arithmetic mean of r and s, so $r < x < s$. That's the conclusion we seek: For any two distinct rational numbers, there exists a third rational number whose value lies between them.

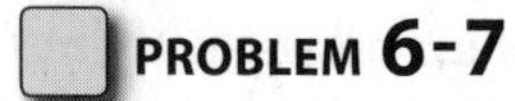

PROBLEM 6-7

Use mathematical induction to show that all natural-number multiples of 0.1 are rational numbers.

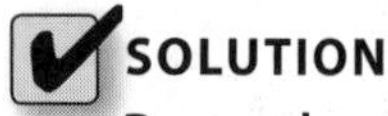

SOLUTION

Remember that we denote the set N of natural numbers as

$$N = \{0, 1, 2, 3, 4, ...\}$$

Therefore, the set M of natural-number multiples of 0.1 is

$$\begin{aligned} M &= \{(0 \times 0.1), (1 \times 0.1), (2 \times 0.1), (3 \times 0.1), (4 \times 0.1), ...\} \\ &= \{0, 0.1, 0.2, 0.3, 0.4, ...\} \end{aligned}$$

The first element of this set, 0, is rational, because we can express it in the form a/b, where a represents an integer and b represents a nonzero natural number. We can simply let $a = 0$ and $b = 1$. That's the easy part of our proof.

Now suppose that $n \times 0.1$ (which we can also denote as $0.1n$) is rational for some unspecified natural number n. Consider the next number in our set M of multiples. That number is $(n + 1) \times 0.1$. According to the rules of arithmetic, we can rearrange it as follows:

$$\begin{aligned} (n + 1) \times 0.1 &= (n \times 0.1) + (1 \times 0.1) \\ &= 0.1n + 0.1 \end{aligned}$$

We know that there exist some integer a and some nonzero natural number b such that $0.1n = a/b$, because we've assumed that $0.1n$ is rational. Therefore, we can rewrite the above expression to obtain

$$\begin{aligned} 0.1n + 0.1 &= a/b + 0.1 \\ &= a/b + 1/10 \end{aligned}$$

Using the arithmetic rule for the sum of two quotients, we can rearrange the above equation to get

$$a/b + 1/10 = (10a + b)/(10b)$$

When we multiply any integer by 10, we get another integer. Therefore, $10a$ represents an integer. The sum of any integer and a nonzero natural number is an integer, so $10a + b$ represents an integer. Ten times any

nonzero natural number equals another nonzero natural number, so the quantity $(10a + b)/(10b)$ equals an integer divided by a nonzero natural number. By definition, it follows that $(10a + b)/(10b)$ represents a rational number. This quantity also happens to equal $a/b + 1/10$, which in turn equals $0.1n + 0.1$, the element immediately after $0.1n$ in the set M.

We've now proved that if any unspecified element of M is rational, then the next element is rational. That, in addition to the proof that the first element in M is rational, is all we need to claim that every element in the set M is rational, based on the principle of mathematical induction.

Cause, Effect, and Implication

When two events occur together, people tend to believe that one event causes the other, especially if the coincidence occurs repeatedly. However, *correlation* does not logically imply *causation*.

Correlation and Causation

Suppose that two phenomena, called X and Y, vary in intensity with time. Figure 6-7 shows a relative graph of the variations in both phenomena, which follow intensity-versus-time curves having similar contours. These curves show *positive correlation*. As X increases, so does Y, in general. As Y decreases, so does X, in general.

Does Fig. 6-7 illustrate a situation in which one phenomenon causes the other? We can't tell by merely observing the graph. Taken all by itself, Fig. 6-7 indicates nothing more than a coincidence. The phenomena X and Y tend to vary along with each other, but the illustration gives us no clue as to the reason.

If there were a few hundred points on each plot, and if the observed phenomena still followed each other as they do in Fig. 6-7, we could make a strong case for the notion that a cause-effect relationship must exist. (How could the correlation be so precise if causation were not involved somehow?) But suppose the graphs represented a "freak scenario"? Suppose the points were selected by someone with a vested interest in the outcome of our analysis?

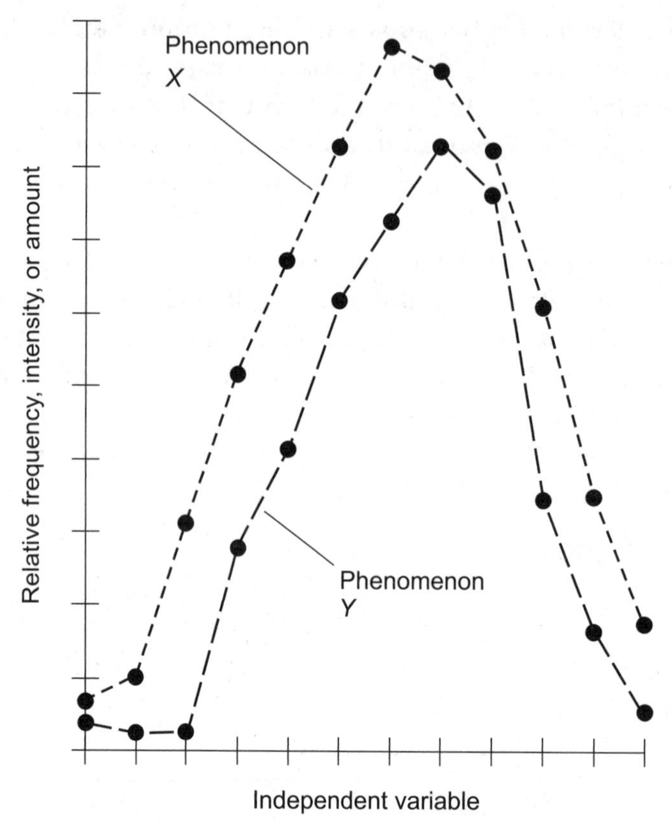

FIGURE 6-7 • The two phenomena *X* and *Y* are correlated, but correlation does not imply causation!

AN EXAMPLE

We can illustrate cause-and-effect relationships with simple diagrams. Figure 6-8A shows a generic cause-and-effect scenario where variations in a phenomenon *X* cause changes in another phenomenon *Y*. (Don't confuse the arrows in this illustration with logical implication arrows!)

Imagine a situation in which the *independent variable*, shown on the horizontal axis in Fig. 6-7, represents the time of day between sunrise and sunset. Suppose that graph *X* shows the relative intensity of sunshine during this time period, while graph *Y* shows the relative temperature over the same period. We might think that the brilliance of the sunshine causes the changes in temperature. A certain time lag appears to exist in the temperature function, but that shouldn't surprise us. The hottest part of the day usually occurs a little later than the time of day when the sun shines with its greatest intensity.

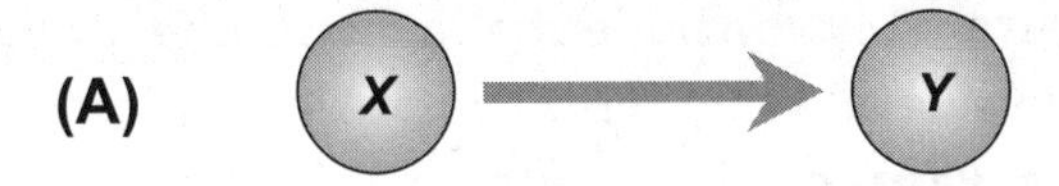

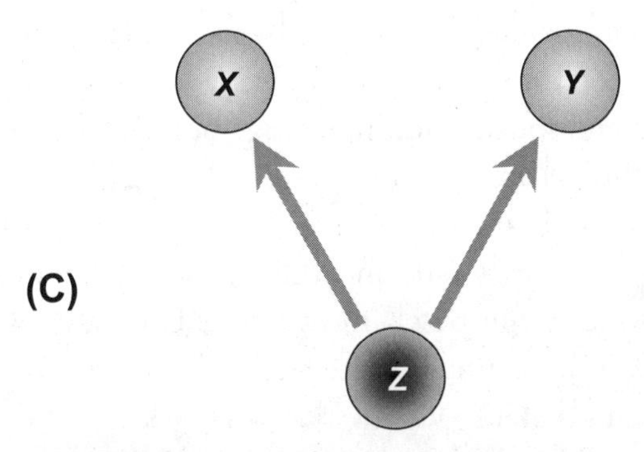

FIGURE 6-8 • At A, *X* causes *Y*. At B, *Y* causes *X*. At C, *Z* causes both *X* and *Y*.

We can imagine a cause-and-effect relationship that works in the other direction, although we might have a difficult time thinking up a good physical explanation. Could changes in temperature cause variations in the intensity of sunlight reaching the earth's surface? At first, that idea seems ridiculous, but keep thinking! Maybe, as the air gets warmer in the particular place for which Fig. 6-7 applies, the clouds dissipate in the atmosphere, allowing more sunlight to reach the surface. In that case, we could argue that *Y* causes *X*.

A SECOND EXAMPLE

Imagine that the horizontal axis in Fig. 6-7 represents 12 different groups of people in a hypothetical medical research survey. Each hash mark on the horizontal axis represents one group. Plot *X* shows the relative number of fatal strokes in a given year for the people in each of the 12 groups; plot *Y* shows the relative average blood pressure levels of the people in the 12 groups during the same year.

Can we reasonably conclude that a cause-effect relationship exists between the value of *X* and the value of *Y* here? Some medical experts might say that variations in *Y* cause, or at least contribute to, observed variations in *X* (Fig. 6-8B). What about the reverse argument? Can fatal strokes cause high blood pressure (*X* causes *Y*)? Of course not. After death, blood pressure loses all relevance!

Complications

The above-described scenarios obviously represent oversimplifications. In real life, events rarely occur with a single clear-cut cause and a single inevitable effect.

The brightness of sunshine does not, all by itself, constitute the only cause of changes in the temperature during the course of a day. A nearby lake or ocean, the wind direction and speed, and the passage of a weather front can all influence the temperature at any given location. We've all seen the weather clear and brighten, along with an abrupt drop in temperature, after a strong cold front passes. The sun comes out, and the air gets dramatically cooler. That observation defies the simplistic claim that bright sun always causes things to heat up, even though the notion is quite reasonable in its "pure" form where all other factors remain constant.

In regards to the blood-pressure-versus-stroke relationship, numerous other factors come into play; even the scientists aren't sure that they know every detail. New discoveries constantly arise in the medical field. Examples of other factors that might play cause-effect roles in the occurrence of fatal strokes include dietary habits, body fat index, diabetes, age, and heredity.

A THIRD EXAMPLE

Suppose that the horizontal axis in Fig. 6-7 represents 12 different groups of people in another medical research survey. Again, each hash mark on the horizontal axis represents one group. Plot *X* illustrates the relative number of heart attacks in a given year for the people in each of the 12 groups; plot *Y* shows the relative average blood cholesterol levels of the people in the 12 groups during the same year. As in the stroke scenario, these are hypothetical graphs. But they're plausible. Medical scientists have demonstrated a correlation between blood cholesterol levels and the frequency of heart attacks.

When doctors and their students first began examining the bodies of people who died of heart attacks in the middle 1900s, they found "lumps" called *plaques* in the arteries. Scientists theorized that plaques cause blood clots that can cut off the circulation to parts of the heart, causing tissue death. The plaques contain cholesterol. Evidently, cholesterol can accumulate inside the arteries. When the scientists saw data showing a correlation between blood cholesterol levels and heart attacks, they theorized that if the level of cholesterol in the blood could be reduced, the likelihood of the person having a heart attack later in life would go down.

Heart specialists began telling their patients to eat fewer cholesterol-containing foods, hoping that this dietary change would reduce blood cholesterol levels. In many cases, that happened. Evidence accumulated that a low-cholesterol diet reduces the likelihood that a person will have a heart attack. There's more than mere correlation here. There's causation, too. Between what variables, and in what directions, does the causation operate?

Let Z represent a high-cholesterol diet. According to contemporary medical-science theory, a cause-and-effect relation exists between this factor and both X and Y. More than a few studies have indicated that, all other things being equal, people who eat lots of cholesterol-rich foods have more heart attacks than people whose diets are cholesterol-lean. Figure 6-8C illustrates this scenario. A cause-and-effect relation exists between Z (lots of cholesterol in the diet) and X (the number of heart attacks); a cause-and-effect relation also exists between Z and Y (the average blood cholesterol level).

A FOURTH EXAMPLE

Let's introduce and identify three new variables. The symbol S represents "stress" (mental or physical hardship), H represents "hereditary background" (genetic makeup), and W represents "working out" (lots of physical activity). Scientists have suggested that cause-and-effect relationships exist between each of these factors and blood cholesterol levels, and between each of these factors and the frequency of heart attacks. Figure 6-9 illustrates several examples.

No matter how nearly the gray arrows in Figs. 6-8 and 6-9 represent true causation in the real world, these relationships never approach the refined status of logical implication in the mathematical sense. Yet, all too often, people would like you to believe that they do. Crafty marketers might try

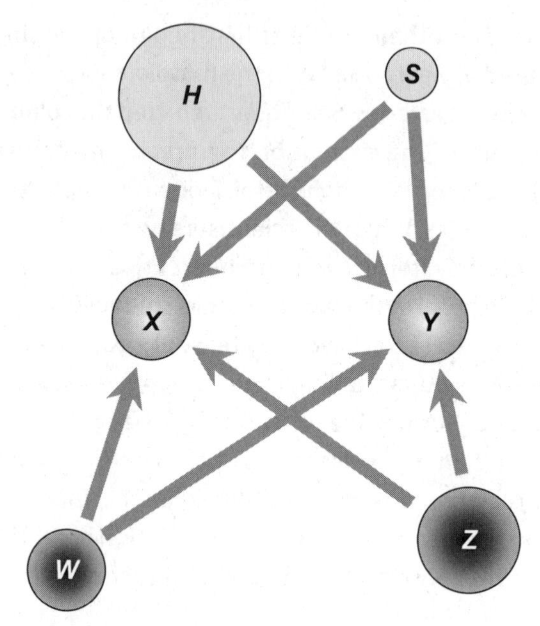

FIGURE 6-9 • A situation in which four different causes *H*, *S*, *W*, and *Z* give rise to two different effects *X* and *Y*.

to get you to buy certain food products, for example, by insinuating (often with the help of dramatic video) that eating cholesterol is as dangerous as jumping off a cliff.

TIP *All of the preceding arguments and examples relate to a single question that we can answer with a truth table. Is logical conjunction equivalent to logical implication? The answer is no; the truth values don't match.* Correlation does not logically imply cause-and-effect!

PROBLEM 6-8

Name some hypothetical cause-and-effect relationships between pairs of the factors in Fig. 6-9, other than those already portrayed. Describe these relationships verbally.

Consider the following propositions. Then ask yourself, "Which of them seem reasonable? Which ones seem possible, but unlikely? Which ones appear impossible or ridiculous?"

- $H \rightarrow S$ (Because of their hereditary backgrounds, some people experience more stress than others.)
- $H \rightarrow Z$ (Because of their hereditary backgrounds, some people consume more cholesterol than others.)
- $H \rightarrow W$ [People with certain hereditary backgrounds work out (exercise) more than people with other hereditary backgrounds.]
- $S \rightarrow H$ (Stress affects people's hereditary backgrounds.)
- $S \rightarrow W$ (Stress affects the extent to which people work out.)
- $S \rightarrow Z$ (Stress affects the amount of cholesterol that people consume.)
- $W \rightarrow H$ (Working out affects a person's hereditary background.)
- $W \rightarrow S$ (Working out affects the amount of stress that people experience.)
- $W \rightarrow Z$ (Working out affects the amount of cholesterol that people consume.)
- $Z \rightarrow H$ (The amount of cholesterol that people consume affects their hereditary backgrounds.)
- $Z \rightarrow S$ (The amount of cholesterol that people consume affects the amount of stress that they experience.)
- $Z \rightarrow W$ (The amount of cholesterol that people consume affects the extent to which they work out.)

QUIZ

You may refer to the text in this chapter while taking this quiz. A good score is at least 8 correct. Answers are in the back of the book.

1. **Suppose we want to prove that there exists no irrational number that we can write down completely in decimal form. Which of the following strategies will most likely work?**
   A. Use the technique of existential quantification along with mathematical induction in an attempt to derive a true statement.
   B. Assume that there is no such number, and then attempt to derive a true statement with deductive reasoning.
   C. Assume that there is such a number, and attempt to derive a contradiction with deductive reasoning.
   D. Use the technique of universal quantification along with deductive reasoning in an attempt to derive a false statement.

2. **We might use mathematical induction to prove the truth of a proposition for**
   A. all of the natural numbers.
   B. all of the negative integers.
   C. all of the integers.
   D. Any of the above.

3. **As we develop a new mathematical system, we might use the process of deduction to**
   A. eliminate ridiculous notions.
   B. offer a convincing argument.
   C. suggest a new theorem.
   D. prove a lemma.

4. **We have learned that correlation does not logically imply causation. In fact, we can demonstrate it as a theorem with a truth table. Which, if any, of the following statements A, B, or C is logically equivalent to this theorem?**
   A. Causation logically implies correlation.
   B. Causation does not logically imply correlation.
   C. The absence of correlation logically implies the absence of causation.
   D. None of the above statements A, B, or C is logically equivalent to the theorem.

5. **A mathematical system is unsound if it leads to**
   A. unnecessary or irrelevant theorems.
   B. unnecessary or irrelevant axioms.
   C. meaningless elementary terms.
   D. a logical contradiction.

6. **Suppose that someone conducts a research study concluding that the number of computers sold per month in the town of Hoodooburg correlates positively with the number of pizzas sold per month in Hoodooburg. From this result, we can logically conclude that**
   A. in Hoodooburg, computer users eat more pizza than noncomputer users.
   B. computer sales in Hoodooburg drives pizza consumption.
   C. pizza consumption in Hoodooburg stimulates computer use.
   D. None of the above.

7. **A strong theorem holds true for**
   A. at least one element of a specified universe.
   B. most of the elements of a specified universe.
   C. every element in a specified universe.
   D. infinitely many elements in a specified universe.

8. **Imagine that we've proved the truth of a proposition in a system of number theory, but we've proven the same proposition false in another, entirely different system of number theory. On this basis alone,**
   A. we don't have a problem.
   B. both systems are unsound.
   C. the system where the proposition holds true is unsound.
   D. the system where the proposition proves false is unsound.

9. **Someone claims that a certain proposition Q holds true for all rational numbers *r*. How might we show that Q is false?**
   A. Assume that Q is false, and then derive a contradiction from that assumption.
   B. Prove that mathematical induction won't work in any attempt to prove the truth of Q.
   C. Prove that Q fails when $r = 0$.
   D. We can't, because we'd have to "prove a negative" for an infinite number of cases, an impossible task.

10. **How can we tell the difference between a lemma and a corollary?**
   A. A lemma follows from the proof of a proposition; a corollary helps us prove a proposition.
   B. A lemma helps us prove a proposition; a corollary follows from the proof of a proposition.
   C. We can assume the truth of a lemma without proof; we can't do that with a corollary.
   D. A lemma can contain one or more elementary terms; a corollary can't contain any.

chapter 7

Boolean Algebra

Boolean algebra constitutes a system of propositional logic in which we represent "truth" as 1 and "falsity" as 0. *Boolean variables* combine to form *boolean expressions* or *boolean equations*. Much of this chapter parallels Chap. 2, but in a different "dialect." In Chap. 2, we thought and wrote like philosophers and mathematicians. Now, we'll think and write like engineers.

CHAPTER OBJECTIVES

In this chapter, you will

- Learn the boolean symbols for logical operations.
- Construct boolean truth tables.
- Define and implement the rules of boolean algebra.
- Redefine common logical laws in boolean terms.
- Prove theorems using boolean algebra.

New Symbols for Old Operations

We represent boolean statements (or sentences) by writing uppercase letters of the alphabet, just as we do in propositional logic. For example, you might say "It's raining" and represent the sentence as R. Your friend might say "It's cold" and represent that statement as C. A third person might say "It will snow tomorrow" and represent it as S. Still another person might say "Tomorrow will be breezy" and represent it as B.

The NOT Operation (−)

When we write a letter to denote a sentence in boolean algebra, we mean to assert that the sentence is true. If John writes C in the above-described situation, he says, in effect, "It's cold." You might also say that C is true if you grew up in Hawaii. But if you live in Alaska, you might say "It's not cold" and denote this declaration by writing the letter C preceded by a negation symbol. In boolean algebra, the minus sign (−) symbolizes the *NOT operation*, so you'd denote "It's not cold" by writing −C.

Now imagine that someone complicates our weather assessment by saying "You are correct to say −C. As a matter of fact, I think it's hot!" Suppose we symbolize the statement "It's hot" as H. In the opinion of any particular person, H implies −C, but that doesn't mean that H is logically equivalent to −C. We can have intermediate conditions between "cold" and "hot," but there exists no meaningful condition between "cold" (C) and "not cold" (−C).

The AND Operation (×)

Suppose that someone says, "It's cold and it's raining." Using the symbols above, we can write this statement as

C AND R

In boolean algebra, we use a multiplication symbol (×) in place of the word AND, so we write the sentence as

$$C \times R$$

Some logicians use an asterisk instead of the conventional multiplication symbol to represent the *AND operation*, so they'll write

C * R

Still other logicians use a raised dot for boolean multiplication, so they'll write

C · R

TIP ***A boolean expression containing one or more AND operations has the value 1 if and only if both (or all) of its components have the value 1. If any component has the value 0, then the whole expression has the value 0.***

The OR Operation (+)

One of your friends says, "It's cold and raining now. The weather experts say it's going to get colder and stay wet, so it might snow tomorrow."

You say, "It will rain or snow tomorrow, depending on the temperature."

Your friend says, "We might see rain and snow together."

You reply, "We might get rain, or we might get snow, or we might get both."

Your friend says, "The weather experts say that we're certain to get precipitation of some sort."

You say, "Water will fall from the sky tomorrow. Maybe it will be liquid, maybe it will be solid, and maybe it will be a combination of both."

Let R represent the sentence "It will rain tomorrow," and let S represent the sentence "It will snow tomorrow." You can write the following statement to summarize the conclusions to which you and your friend have come:

S OR R

This statement provides us with an example of the *OR operation* in action. In boolean algebra, we use the addition symbol to represent OR, so we can write the above sentence as

S + R

TIP ***A boolean expression in which both (or all) of the components are joined by OR operations has the value 1 if and only if at least one component has the value 1. A boolean expression made up of OR operations has the value 0 if and only if both (or all) of the components have the value 0.***

TIP ***Have you noticed that we've discussed the logical*** **inclusive OR** ***operation here? In digital electronics and computer engineering, you'll occasionally encounter another boolean operation called*** **exclusive OR** ***(XOR), in which a two-variable expression has the value 1 if and only if one variable has the value 1 while the other value has the value 0. The boolean XOR operation is the equivalent of the logical connector*** **EITHER/OR.**

Boolean Implication (⇒)

You and your friends must decide if you should get ready for a snowstorm tomorrow, or whether you'll have to contend with nothing worse than rain.

You ask, "Does the weather forecast say anything about snow?"

Your friend says, "Not directly. The weather experts say 'Precipitation will fall steadily through tomorrow night, and the temperature will get colder tomorrow.' I looked at my outdoor thermometer when I heard that, and the thermometer registered slightly above freezing."

You say, "If there is precipitation, and if it gets colder, then it will snow."

Your friend says, "Of course."

You add, "Unless we get an ice storm."

Your friend insists, "That won't happen."

You say, "Let's rule out the ice-storm scenario. That means that if there is precipitation tomorrow, and if it's colder tomorrow than today, then it will snow tomorrow."

Let P represent the sentence "Precipitation will fall tomorrow." Let S represent the sentence "Snow will fall tomorrow." Let C represent the sentence "It will be colder tomorrow than it is today." In the foregoing conversation, you and your friend have arrived at the compound statement

IF (P AND C), THEN S

The parentheses indicate how the variables are grouped. Operations within parentheses are always done before operations outside them. You can also write the above statement as

(P AND C) IMPLIES S

In boolean algebra, "X implies Y" is the equivalent of saying "If X, then Y." If X is true, then we *know* that Y is true. If event X occurs, then event Y is *certain*. If we observe phenomenon X, then (if we look hard enough) we *will* see phenomenon Y. Symbolically, we can write the above compound statement as

$$(P \times C) \Rightarrow S$$

The double-shafted, right-pointing arrow represents *boolean implication*, also known as the *IF/THEN operation*. The "implying" sentence (to the left of the arrow) is called the *antecedent*. In this case, the antecedent is $(P \times C)$. The "implied" sentence (to the right of the arrow) is called the *consequent*. Here, the consequent is S.

Boolean Equivalence (=)

Suppose one of your friends continues the conversation by saying, "If it snows tomorrow, then there will be precipitation and it will be colder."

For a moment you hesitate, because this isn't the way you'd usually think about this sort of scenario. But you have to agree, "That's true. It sounds strange, but it's totally logical." Your friend has made the implication

$$S \Rightarrow (P \times C)$$

You and your friend have already concluded that

$$(P \times C) \Rightarrow S$$

When boolean implication operates in both directions, you have an instance of *boolean equivalence*. You can write the combination of the preceding two statements as

$(P \times C)$ IF AND ONLY IF S

Mathematicians often shorten IF AND ONLY IF to IFF, so they might write

$(P \times C)$ IFF S

The symbol for boolean equivalence, also called *boolean equality*, is an equals sign (=), so we'd write the above conclusion as

$$(P \times C) = S$$

TIP ***Don't confuse this application of the equals sign with use of the same symbol to assign truth values to statements in propositional logic!***

PROBLEM 7-1

Provide an example of a situation in which logical implication holds in one direction but not in the other.

SOLUTION

Consider the sentence "If it's overcast, then there are clouds in the sky." This statement is always true, because "overcast" means "completely cloud-covered." Suppose we let O represent "It's overcast" and K represent "There are clouds in the sky." Then we can write

$$O \Rightarrow K$$

If we reverse the order of the variables, we get

$$K \Rightarrow O$$

This boolean expression translates to the statement, "If there are clouds in the sky, then it's overcast." That's not necessarily true. We've all seen days or nights in which clouds were visible, but they did not completely fill the sky.

Truth Tables, Boolean Style

As you learned earlier in this book, a truth table shows all of the possible combinations of truth values for the variables in a proposition. We write the values for the individual variables in vertical columns at the left. We show the values for boolean expressions, as they arise from combinations of *boolean variables* (unbreakable sentences), in horizontal rows.

Truth Table for Boolean Negation (NOT)

The simplest truth table is the one for negation, which operates on a single variable. Table 7-1 shows how boolean negation works for a variable X.

TABLE 7-1 Truth table for boolean negation, also known as the NOT operation.

X	–X
0	1
1	0

Truth Table for Boolean Multiplication (AND)

Consider two variables X and Y. Boolean multiplication $(X \times Y)$ produces results as shown in Table 7-2. The resultant equals 1 if and only if both variables have value 1. If either or both variables equal 0, then the boolean product equals 0.

TABLE 7-2 Truth table for boolean multiplication, also known as the AND operation.

X	Y	$X \times Y$
0	0	0
0	1	0
1	0	0
1	1	1

Truth Table for Boolean Addition (OR)

Table 7-3 shows the truth table for the boolean sum of two variables $(X + Y)$. The resultant has the value 1 when either or both variables equal 1. If both variables equal 0, then the boolean sum equals 0. Remember that a boolean sum represents the inclusive OR operation, not the exclusive OR operation!

TABLE 7-3 Truth table for boolean addition, also known as the OR operation.

X	Y	X + Y
0	0	0
0	1	1
1	0	1
1	1	1

Truth Table for Boolean Implication (IF/THEN)

A case of boolean implication is valid (i.e., it has truth value 1), except when the antecedent equals 1 and the consequent equals 0. Table 7-4 shows the truth values for this operation.

TABLE 7-4 Truth table for boolean implication, also known as the IF/THEN operation.

X	Y	$X \Rightarrow Y$
0	0	1
0	1	1
1	0	0
1	1	1

TIP ***Note that while boolean multiplication and addition are* commutative *(the order of the operation doesn't matter), boolean implication is not commutative (the order matters).***

Truth Table for Boolean Equality (IFF)

If X and Y constitute logical variables, then X = Y is a valid statement if and only if both variables have the same value. If the truth values of X and Y differ, then X = Y is not valid. Table 7-5 shows all possible conditions for the two-variable case.

TABLE 7-5 Truth table for boolean equality, also called the IFF operation.

X	Y	X = Y
0	0	1
0	1	0
1	0	0
1	1	1

PROBLEM 7-2

Provide a verbal example of a boolean implication that's obviously invalid.

SOLUTION

Let X represent the sentence, "You see a thunderstorm in the distance." Let Y represent the sentence, "A thunderstorm is coming toward you." Now consider the following statement:

$$X \Rightarrow Y$$

Imagine that you can see a thunderstorm several kilometers away as it drifts along from west to east with the prevailing winds. In this scenario, $X = 1$. Suppose you're located west of the storm, so it's moving away from you. That means $Y = 0$. Because the antecedent equals 1 and the consequent equals 0, the boolean implication has truth value 0. The fact that you can see a storm doesn't logically imply that it's approaching you.

PROBLEM 7-3

Derive the truth table for boolean equality based on the truth tables for boolean multiplication and boolean implication.

SOLUTION

Remember that $X = Y$ means the same thing as $(X \Rightarrow Y) \times (Y \Rightarrow X)$. Based on this fact, we can build up $X = Y$ in steps, as shown in Table 7-6 from left to right. The four possible combinations of truth values for sentences X and Y

TABLE 7-6 Solution to Problem 7-3.

X	Y	$X \Rightarrow Y$	$Y \Rightarrow X$	$X = Y$
0	0	1	1	1
0	1	1	0	0
1	0	0	1	0
1	1	1	1	1

appear in the first (left-most) and second columns. The truth values for the statement $X \Rightarrow Y$ appear in the third column, and the truth values for the statement $Y \Rightarrow X$ appear in the fourth column. The truth values in the fifth or right-most column ($X = Y$) represent the boolean products of the truth values in the third and fourth columns.

PROBLEM 7-4

Consider an operation called NAND, which consists of the AND operation acting on two variables, followed by negation of the result. Suppose that we denote this operation as a multiplication sign with a circle around it (⊗). Define this operation symbolically, and write down a truth table for it.

SOLUTION

Suppose that X and Y are logical variables. Symbolizing the NAND operation as ⊗, we can write

$$X \otimes Y = -(X \times Y)$$

Table 7-7 shows how we derive the truth values for this operation in terms of boolean multiplication and negation.

TABLE 7-7 Solution to Problem 7-4.

X	Y	X × Y	X ⊗ Y
0	0	0	1
0	1	0	1
1	0	0	1
1	1	1	0

PROBLEM 7-5

Consider an operation called NOR, which consists of the inclusive OR operation acting on two variables, followed by negation of the result. Suppose that we symbolize this operation by writing an addition (plus) sign with a circle around it (⊕). Define this operation symbolically, and write down a truth table for it.

SOLUTION

Suppose that X and Y are logical variables. Symbolizing the NOR operation as ⊕, we can write

$$X \oplus Y = -(X + Y)$$

Table 7-8 shows how we derive the truth values for this operation in terms of boolean addition and negation.

TABLE 7-8 Solution to Problem 7-5.

X	Y	X + Y	X ⊕ Y
0	0	0	1
0	1	1	0
1	0	1	0
1	1	1	0

PROBLEM 7-6

Consider the exclusive OR operation (called XOR) that we defined earlier in this chapter. Suppose that we symbolize this operation as a plus sign with a minus sign underneath (±). Write down a truth table for this two-variable operation.

SOLUTION

Suppose that X and Y are logical variables. If we symbolize the XOR operation as ±, then we have

$$X \pm Y = 1 \text{ when } X \neq Y$$

and

$$X \pm Y = 0 \text{ when } X = Y$$

Table 7-9 shows the values of the XOR operation.

TABLE 7-9 Solution to Problem 7-6.

X	Y	X ± Y
0	0	0
0	1	1
1	0	1
1	1	0

PROBLEM 7-7

Consider an operation called XNOR (or, alternatively, NXOR), which consists of the XOR operation acting on two variables, followed by negation of the result. Suppose that we symbolize this operation by writing an old-fashioned division sign (÷). Define this operation symbolically, and write down a truth table for it, as applicable to two variables.

SOLUTION

Let X and Y be logical variables. If we symbolize the XOR operation as ±, and if we symbolize the XNOR operation as ÷, then

$$X \div Y = -(X \pm Y)$$

Table 7-10 shows how we derive the truth values for this operation in terms of XOR followed by negation.

TABLE 7-10 Solution to Problem 7-7.

X	Y	X ± Y	X ÷ Y
0	0	0	1
0	1	1	0
1	0	1	0
1	1	0	1

Basic Boolean Laws

Boolean operations obey specific *mathematical laws*. Remember that, in order to "qualify" as a mathematical law, a generalized statement must hold true in all possible instances. Some fundamental boolean laws follow. Try to memorize them! You should find that task rather easy, because these rules "mirror" the rules for propositional logic.

Precedence

When you read or construct a complicated logical statement in boolean algebra, you can use the same grouping symbols that you would use in ordinary algebra or in propositional logic.

You should always do, or expect your reader to perform, operations within parentheses before operations outside them. If you have a statement that involves multilevel combinations of sentences (so-called *nesting of operations*), then you can use ordinary round parentheses first, then square parentheses (called *brackets*) outside of the round ones, and then, if necessary, curly parentheses (called *braces*) outside of the square ones. Alternatively, you can use sets of plain parentheses inside each other.

Whatever grouping symbols you decide to use, a complete boolean statement or equation must always contain the same number of left-hand and right-hand grouping symbols of each type.

If you find no grouping symbols in a boolean expression, you should work out all instances of negation first. Then you should work out all the products, then all the sums, then the implications, and finally the equivalences. For example, consider the boolean statement

$$A \times -B + C \Rightarrow D$$

Using parentheses, brackets, and braces to clarify this statement according to the rules of precedence, you can write

$$\{[A \times (-B)] + C\} \Rightarrow D$$

If you want to use only conventional parentheses, you can write

$$((A \times (-B)) + C) \Rightarrow D$$

Contradiction

A *contradiction* always results in a false truth value (logic 0). Lay people sometimes express this principle by saying, "From a contradiction, anything follows." Unfortunately, some people abuse that expression to come up with ridiculous statements such as, "If I am a robot and I am not a robot, then I am the Czar of Saturn." In a purely boolean context, this principle merely tells us that if X is a logical variable, then

$$X \times -X \Rightarrow 0$$

Law of Double Negation

The negation of a negation equals the original expression. That is, for any boolean variable X,

$$-(-X) = X$$

Commutative Laws

Boolean multiplication and addition both obey commutative principles, meaning that they work in either direction. A boolean product or sum has the same value regardless of the order in which we do the operations. If X and Y are boolean variables, then

$$X \times Y = Y \times X$$

and

$$X + Y = Y + X$$

Associative Laws

With a product or sum of three boolean variables, it doesn't matter how you group them. Suppose you have the expression

$$X \times Y \times Z$$

You can consider $X \times Y$ as a single boolean variable and multiply it by Z on the right, or you can consider $Y \times Z$ as a single variable and multiply it by X on the left. Either way, the results are equal. Therefore,

$$(X \times Y) \times Z = X \times (Y \times Z)$$

A similar law holds for boolean sums. You can write that principle as

$$(X + Y) + Z = X + (Y + Z)$$

Engineers and scientists call these rules the *associative law for boolean multiplication* and the *associative law for boolean addition*, respectively.

TIP ***You must use care when applying the associative laws in boolean algebra. All the operations in the expression must be of the same type (either all multiplication or all addition). If a boolean expression contains a product and a sum, you can't change the grouping and then expect to get the same truth value in all possible cases. For example, the statement***

$$(X \times Y) + Z$$

is* not logically equivalent *to the statement

$$X \times (Y + Z)$$

Law of the Contrapositive

When one boolean variable implies another, you can't reverse the sense of the implication and expect the result to remain valid. It is not always true that if $X \Rightarrow Y$, then $Y \Rightarrow X$. However, if you negate both variables and reverse the sense of the implication, you'll always get another valid proposition. We call this principle the *law of the contrapositive*. For any two boolean variables X and Y, we can always say that

$$(X \Rightarrow Y) = (-Y \Rightarrow -X)$$

DeMorgan's Law for Products

If we negate the boolean product of two variables, we can write the result as the boolean sum of the negations of the original variables. We call this rule *DeMorgan's law for products*. Expressed symbolically, if X and Y are boolean variables, then

$$-(X \times Y) = (-X) + (-Y)$$

DeMorgan's Law for Sums

If we negate the boolean sum of two variables, we can write the result as the boolean product of the negations of the original variables. We call this rule *DeMorgan's law for sums*. Expressed symbolically, if X and Y are boolean variables, then

$$-(X + Y) = (-X) \times (-Y)$$

Distributive Law

A specific relationship exists between boolean multiplication and addition, known as the *distributive law*. It works in the same way the *distributive principle* in ordinary arithmetic or algebra, which states that if a and b represent any two numbers, then

$$a(b + c) = ab + ac$$

Remember that boolean addition is the equivalent of logical disjunction (the inclusive-OR operation), and boolean multiplication is the equivalent of logical conjunction (the AND operation). If X, Y, and Z are boolean variables, then

$$X \times (Y + Z) = (X \times Y) + (X \times Z)$$

PROBLEM 7-8

Use words to illustrate a "real-life" example of the law of the contrapositive.

SOLUTION

Let H represent the sentence, "Helen is a human." Let M represent the sentence, "Helen is a mammal." Now suppose that we have the boolean equation

$$H \Rightarrow M$$

We can read this equation out loud by saying, "If Helen is a human, then Helen is a mammal." That statement holds true no matter who (or what) Helen happens to be. When we apply the law of the contrapositive, we obtain the boolean equation

$$-M \Rightarrow -H$$

When we put this equation in words, we get "If Helen is not a mammal, then Helen is not human." That's also true. None of us will ever encounter a nonmammalian human (no matter how much we might sometimes want to call certain people "reptiles" or "fish" or whatever)!

PROBLEM 7-9

Construct a pair of truth tables illustrating the validity of DeMorgan's law for products.

SOLUTION

Let X and Y be boolean variables. Tables 7-11A and 7-11B demonstrate that

$$-(X \times Y) = (-X) + (-Y)$$

All the truth values in the far-right column of Table A equal the corresponding truth values in the far-right column of Table B, so we know that the boolean variables at the tops of those columns always have equal value.

TABLE 7-11 Solution to Problem 7-9.

A

X	Y	X × Y	−(X × Y)
0	0	0	1
0	1	0	1
1	0	0	1
1	1	1	0

B

X	Y	−X	−Y	(−X + −Y)
0	0	1	1	1
0	1	1	0	1
1	0	0	1	1
1	1	0	0	0

PROBLEM 7-10

Construct a pair of truth tables showing the validity of DeMorgan's law for sums.

SOLUTION

Let X and Y be boolean variables. Tables 7-12A and 7-12B demonstrate that

$$-(X + Y) = (-X) \times (-Y)$$

All the truth values in the far-right column of Table A equal the corresponding truth values in the far-right column of Table B, so we know that the boolean variables at the tops of those columns always have equal value.

PROBLEM 7-11

Construct a pair of truth tables that demonstrates the validity of the distributive law.

TABLE 7-12 Solution to Problem 7-10.

A

X	Y	X + Y	−(X + Y)
0	0	0	1
0	1	1	0
1	0	1	0
1	1	1	0

B

X	Y	−X	−Y	(−X × −Y)
0	0	1	1	1
0	1	1	0	0
1	0	0	1	0
1	1	0	0	0

SOLUTION

Let X, Y, and Z be boolean variables. Tables 7-13A and 7-13B demonstrate that

$$X \times (Y + Z) = (X \times Y) + (X \times Z)$$

All the truth values in the far-right column of Table A are identical to the corresponding truth values in the far-right column of Table B, so we know that the boolean variables at the tops of those columns have equal value.

PROBLEM 7-12

A *boolean function* comprises a combination of operations on two or more variables (called *inputs*), producing a single resultant variable (or *output*) with a set of values that depends on the values of the inputs. Write down a boolean function for five inputs A, B, C, D, and E, such that the output Z equals 0 when all the inputs equal 1, but Z = 1 if any of the inputs equal 0.

TABLE 7-13 Solution to Problem 7-11.

A				
X	Y	Z	Y + Z	X × (Y + Z)
0	0	0	0	0
0	0	1	1	0
0	1	0	1	0
0	1	1	1	0
1	0	0	0	0
1	0	1	1	1
1	1	0	1	1
1	1	1	1	1

B					
X	Y	Z	X × Y	X × Z	(X × Y) + (X × Z)
0	0	0	0	0	0
0	0	1	0	0	0
0	1	0	0	0	0
0	1	1	0	0	0
1	0	0	0	0	0
1	0	1	0	1	1
1	1	0	1	0	1
1	1	1	1	1	1

SOLUTION

Consider the following boolean equation, in which we obtain the output by multiplying all the inputs and then negating, as follows:

$$Z = -(A \times B \times C \times D \times E)$$

In this equation, Z = 0 if and only if all of the input variables equal 1. If one or more of the input variables equals 0, then Z = 1. You can verify this fact by constructing a truth table. You'll need a large sheet of lined paper (or a large computer display) if you want to see the whole table at once, because the table will have 32 rows!

PROBLEM 7-13

Write down another boolean function, different from the one you found in the solution to Problem 7-12, for five variables A, B, C, D, and E, such that the output Z equals 0 when all the inputs equal 1, but such that Z = 1 if any of the inputs equal 0.

SOLUTION

Consider the boolean equation that you get by adding up the negations of all the inputs, as follows:

$$Z = (-A) + (-B) + (-C) + (-D) + (-E)$$

In this equation, Z = 0 if and only if all of the inputs, A through E, equal 1. If one or more of the inputs, A through E, equal 0, then Z = 1. You can verify this fact using a truth table, although it will be 32 rows long.

TIP ***Have you noticed that Problem 7-13 and its solution, taken along with Problem 7-12 and its solution, demonstrate the fact that DeMorgan's law holds for products of five variables?***

QUIZ

You may refer to the text in this chapter while taking this quiz. A good score is at least 8 correct. Answers are in the back of the book.

1. **The boolean sum of 13 boolean variables equals 1 if**
   A. at least one of the propositions has the value 1.
   B. at least 7 of the propositions have the value 1.
   C. all of the propositions have the value 1.
   D. all of the propositions have the same value.

2. **The boolean product of 13 boolean variables equals 1 if**
   A. at least one of the propositions has the value 1.
   B. at least 7 of the propositions have the value 1.
   C. all of the propositions have the value 1.
   D. all of the propositions have the same value.

3. **A boolean implication has the value 1**
   A. if and only if the left-hand variable equals 1.
   B. if and only if the left-hand variable equals 0.
   C. except when the left-hand variable equals 0 and the right-hand variable equals 1.
   D. except when the left-hand variable equals 1 and the right-hand variable equals 0.

4. **How many possible combinations of truth values exist for a set of five boolean variables, each of which can attain either the value 1 or the value 0?**
   A. 8
   B. 16
   C. 25
   D. 32

5. **In boolean algebra, the logical connector "if and only if" is commonly symbolized as**
   A. a plus sign.
   B. a double-shafted arrow pointing to the right.
   C. a forward slash.
   D. an equals sign.

6. **Suppose I claim that a certain boolean equation holds true in general. You assume that at least one case exists where my equation does not work, and then you derive a contradiction based on that assumption. You've proven**
   A. that from a contradiction, anything follows.
   B. that my equation constitutes a boolean law.
   C. DeMorgan's laws.
   D. the law of the contrapositive.

TABLE 7-14 Truth table for Quiz Question 7.

X	Y	Z	$X \times Z$	$(X \times Z) \Rightarrow Y$
0	0	0	0	1
0	0	1	0	1
0	1	0	0	0
0	1	1	0	0
1	0	0	0	1
1	0	1	1	0
1	1	0	0	0
1	1	1	1	1

7. Look at Table 7-14. If something is wrong, how can we change the table to make it valid?

A. Nothing is wrong.

B. Something is wrong. We can make the table valid by changing the double-shafted arrow in the far-right header to an equals sign.

C. Something is wrong. The boolean statement $(X \times Z) \Rightarrow Y$ is meaningless. We can't do anything to make the table valid.

D. Something is wrong. We can make the table valid by reversing all of the truth values in the far-right column.

8. Look at Table 7-15. If something is wrong, how can we change the table to make it valid?

A. Nothing is wrong.

B. Something is wrong. We can make the table valid by changing the plus sign in the statement at the top of the far-right column to an equals sign.

C. Something is wrong. The boolean statement $(X \times Y) + Z$ is meaningless. We can't do anything to make the table valid.

D. Something is wrong. We can make the table valid by reversing all of the truth values in the far-right column.

TABLE 7-15 Truth table for Quiz Question 8.

X	Y	Z	$X \times Y$	$(X \times Y) + Z$
0	0	0	0	1
0	0	1	0	0
0	1	0	0	1
0	1	1	0	0
1	0	0	0	1
1	0	1	0	0
1	1	0	1	0
1	1	1	1	0

9. Consider the boolean statement $-Y \Rightarrow -X$. Which of the following statements is logically equivalent to it?

A. $X \Rightarrow -Y$
B. $-X \Rightarrow -Y$
C. $X \Rightarrow Y$
D. $X \Rightarrow -Y + X$

10. Consider the boolean statement $(-X) + (-Y)$. Which of the following statements is logically equivalent to it?

A. $-(Y + X)$
B. $-(Y \times X)$
C. $-(Y \Rightarrow X)$
D. $Y + (-X)$

chapter 8

The Logic of Sets

Logic gives rise to an important branch of mathematics known as *set theory*. Mathematicians define a *set* as a collection or group of objects called *elements* or *members*. An element of a set can be anything real or imagined, concrete or abstract.

CHAPTER OBJECTIVES

In this chapter, you will

- Define and symbolize sets.
- See how sets relate to each other.
- Draw diagrams to illustrate set behavior.
- Learn basic set operations.
- Compare set operations with logical operations.

Set Fundamentals

If you have a set of a dozen eggs, you have something more than mere eggs. You have the right to claim that those eggs all belong to a specific group. If you think about the group of all people in the state of South Dakota, you're thinking about something more than people. You're thinking about a specific set of people.

To Belong, or Not to Belong

If we want to call some entity x an element of a defined set A, then we write

$$x \in A$$

The "lazy pitchfork" symbol means "is an element of." We can also say that x "belongs" to set A, or that x "is in" set A. If some other entity y is *not* an element of set A, then we write

$$y \notin A$$

An element constitutes a "smallest possible piece" that can exist in any set. We can't break an element down into anything smaller and have it remain a legitimate member of the original set. This concept becomes important whenever we have a set that contains another set as one of its elements.

Listing the Elements

When we list the elements of a set, we enclose the list in "curly brackets," technically known as *braces*. The order of the list doesn't matter. Repetition in the list doesn't make any difference either. The following sets are all identical:

$$\{1, 2, 3\}$$
$$\{3, 2, 1\}$$
$$\{1, 3, 3, 2, 1\}$$
$$\{1, 2, 3, 1, 2, 3, 1, 2, 3, \ldots\}$$

The *ellipsis* (string of three dots) means that the list goes on forever in the pattern shown. Now look at this example of a set with five elements:

$$S = \{2, 4, 6, 8, 10\}$$

Whether the elements of S are numbers or numerals depends on the context. Usually, when we see a set with numerals listed like this, the author means to define the set containing the numerical values that those numerals represent. Here's another example of a set with five elements:

$$P = \{\text{Mercury, Venus, Earth, Mars, Jupiter}\}$$

We can reasonably assume that the elements of this set are the first five planets in our solar system, not the words representing those planets!

The Empty Set

A set can exist even if it contains no elements. When we have a set lacking elements altogether, we call it the *empty set* or the *null set*. We can symbolize the null set by writing two braces facing each other with a space between:

$$\{\ \}$$

We can also denote the null set as a circle with a forward slash through it:

$$\varnothing$$

You might ask, "How can a set have no elements? That would be like a club with no members." That's exactly right! If all the members of the Pingoville Ping-Pong club quit today and no new members join, the club still exists as long as it has a charter and by-laws. The set of members of the Pingoville Ping-Pong Club might be empty, but it's a legitimate set as long as someone says that the club exists.

Finite or Infinite?

Mathematicians categorize sets as either *finite* or *infinite*. We can name all of the elements of a finite set if we have enough time. This includes the null set. We can utter the words "This set has no elements," and we've named all the elements of the null set. We cannot name all the elements of an infinite set, no matter how much time we have. Nevertheless, we can sometimes write an "implied list" of the elements of an infinite set that describes that set in unambiguous, clear terms. When we encounter a set like this, we call it *denumerably infinite*. Consider the following infinite set:

$$W = \{0, 1, 2, 3, 4, 5, \ldots\}$$

We recognize W as the set of *nonnegative integers*, also called *whole numbers*. We know whether or not something constitutes an element of set W, even if we don't see it in the above "short list," and even if we couldn't reach it if we started to scribble down the "long list" right now and kept at it for years. We can tell straightaway which of the following numbers belong to W and which do not:

12

1/2

23

100/3

78,883,505

356.75

90,120,801,000,000,000

–65,457,333

The first, third, fifth, and seventh numbers belong to W, but the second, fourth, sixth, and eighth numbers don't.

TIP ***Sometimes, we'll encounter an infinite set that we cannot define by means of any list, even an "implied list." The set of real numbers falls into this category. So does the set of irrational numbers. Theorists call sets of this sort*** **nondenumerably infinite** ***or simply*** **nondenumerable.**

Sets within Sets

A set can act as an element of another set. Remember again, anything can serve as a member of a set! When you allow sets to compose members of other sets, you get end up with sets that get confusing! Following are some examples, in increasing order of strangeness:

$$\{1, 2, \{3, 4, 5\}\}$$

$$\{1, \{2, \{3 ,4, 5\}\}\}$$

$$\{1, \{2, \{3, \{4, 5\}\}\}\}$$

$$\{1, \{2, \{3, \{4, \{5\}\}\}\}\}$$

TIP ***An "inner" or "member" set can sometimes contain more elements than the set to which it belongs. Consider the following example:***

{1, 2, {3, 4, 5, 6, 7, 8}}

In this case, the "main" set has three elements, one of which is a set with six elements.

Here, the "main" set has three elements, one of which is a set with six elements.

PROBLEM 8-1

Name a set that's an element of itself.

SOLUTION

The set of all abstract ideas is abstract idea, so it's an element of itself.

PROBLEM 8-2

Build up an infinite set of sets, starting with the null set.

SOLUTION

Consider the set containing the null set, that is, {∅}. This set is a mathematical object, but it's not the null set. It's the sole element of {{∅}}. Now imagine the following "top-down" list of sets, in which any given set acts as the sole element of the next one:

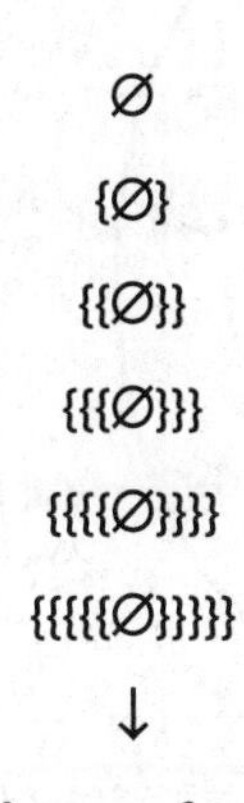

and so on, forever

Venn Diagrams

Relationships among sets of similar objects lend themselves to a special type of illustration called the *Venn diagram*, in which sets appear as groups of points or geometric figures.

People and Numbers

Figure 8-1 illustrates a simple Venn diagram. The large, heavy rectangle represents the set of all things that can exist, whether real or imaginary (including all possible sets). We call this infinite set the *universal set* or the *universe*. In Fig. 8-1, we see three finite sets and two infinite sets within the universe. Note how the objects overlap, lie within one another, or are entirely separated from one another.

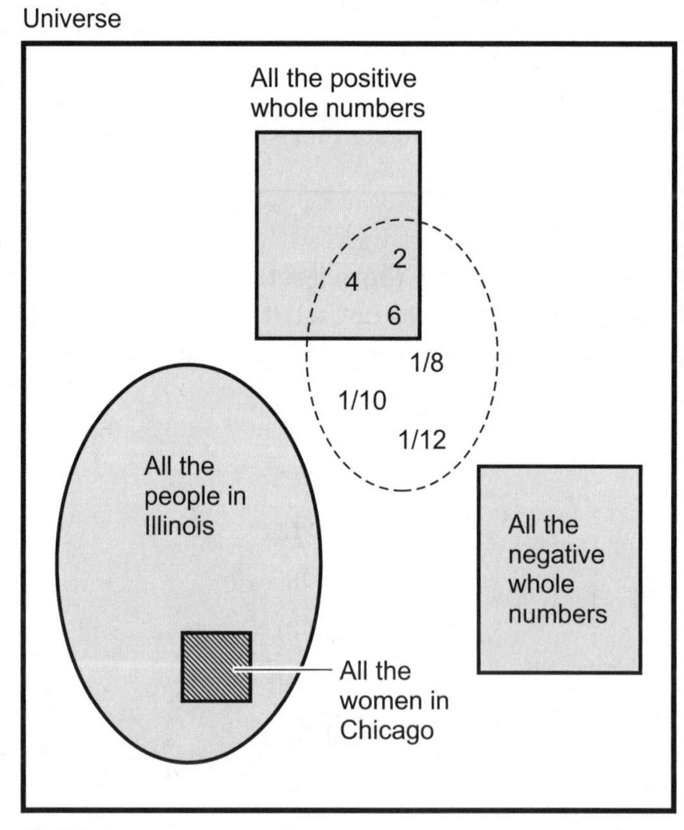

FIGURE 8-1 • A Venn diagram showing the set of all sets (the universe), along with a few specific sets within it.

All the women in Chicago are people in Illinois, but there are plenty of people in Illinois who aren't people in Chicago. The numbers 2, 4, and 6 are positive whole numbers, but there are lots of positive whole numbers different from 2, 4, or 6. The sets of positive and negative whole numbers are entirely separate, even though both sets are infinite. None of the positive or negative whole numbers is a person in Illinois, and no person in Illinois is a number (except according to the government, maybe).

Subsets

When all the elements of a set also belong to a second set, we call the first set a *subset* of the second set. If we encounter two sets A and B, such that every element of A also belongs to B, then A is a subset of B. We symbolize that fact as

$$A \subseteq B$$

Figure 8-1 shows that the set of all the women in Chicago forms a subset of the set of all the people in Illinois. We indicate that fact (in this example, anyhow) by drawing a hatched square inside a shaded oval. Figure 8-1 also shows that the set {2, 4, 6} constitutes a subset of the set of positive whole numbers. We express that fact by placing the numerals 2, 4, and 6 inside the rectangle representing the positive whole numbers. All five of the figures inside the large, heavy rectangle portray subsets of the universe. Any set we can imagine, no matter how large, small, or strange, whether finite or infinite, forms a subset of the universe. A set is always a subset of itself, too.

Proper Subsets

Often, a subset represents part, but not all, of the "main set." In a situation like that, we call the smaller set a *proper subset* of the larger set. In the scenario of Fig. 8-1, the set of all women in Chicago constitutes a proper subset of the set of all people in Illinois. The set {2, 4, 6} is a proper subset of the set of positive whole numbers. All five of the sets inside the large, outermost rectangle are proper subsets of the universe. When a certain set C is a proper subset of another set D, we write this fact as

$$C \subset D$$

PROBLEM 8-3

Name a set that's a subset of every possible set.

SOLUTION

The null set is a subset of any set we care to imagine. If we have a set *A* with known elements, we can add nothing to its roster of members, and we'll always get the same set *A*. The null set is a subset of itself, although not a *proper* subset of itself. Here's an example. If you let the written word "nothing" actually stand for nothing, then

$$\varnothing = \{\,\} = \{\text{nothing}\}$$

and

$$\{\text{nothing}\} \subseteq \{\text{nothing}, 1, 2, 3\}$$

so therefore

$$\varnothing \subseteq \{1, 2, 3\}$$

but

$$\{\text{nothing}\} \not\subset \{\text{nothing}\}$$

Here, the symbol $\not\subset$ means "is not a proper subset of."

TIP *Keep in mind that a* **subset** *does not constitute the same thing as a* **set element**. *The null set* **contains** *nothing, but the null set is not* **itself** *nothing.*

Congruent Sets

Once in awhile, you'll encounter two sets that have different denotations or expressions, but you discover that they're identical when you scrutinize them. Consider the following:

$$E = \{1, 2, 3, 4, 5, \ldots\}$$

and

$$F = \{7/7, 14/7, 21/7, 28/7, 35/7, \ldots\}$$

At first glance, these two sets look completely different. But if you think of their elements as numbers (not as symbols representing numbers), you can see that both lists represent the same set. You know this because

$$7/7 = 1$$
$$14/7 = 2$$
$$21/7 = 3$$
$$28/7 = 4$$
$$35/7 = 5$$
$$\downarrow$$
and so on, forever

Every element in set E has exactly one "mate" in set F, and every element in set F has exactly one "mate" in set E. We can pair off the elements of E and F in a *one-to-one correspondence*.

When two sets contain identical elements, and we can pair off the elements of one set in a one-to-one correspondence with the elements in the other, we call the two sets *congruent sets*. Some mathematicians call them *equal sets* or *coincident sets*. In the above situation, we can write

$$E = F$$

Once in awhile, an author will use a three-barred equals sign to indicate congruence between sets, as follows:

$$E \equiv F$$

Disjoint Sets

When two sets have no elements in common, we call them *disjoint sets*. Here's an example of two disjoint sets of numbers:

$$G = \{1, 2, 3, 4\}$$

and

$$H = \{5, 6, 7, 8\}$$

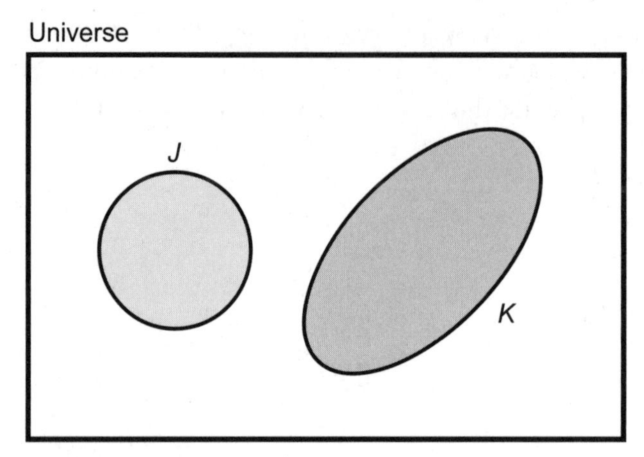

FIGURE 8-2 • Two disjoint sets, *J* and *K*. They have *no* elements in common.

Both of these sets happen to be finite, but infinite sets can also be disjoint. Consider the set of all the even whole numbers and the set of all the odd whole numbers:

$$W_{\text{even}} = \{0, 2, 4, 6, 8, \ldots\}$$

and

$$W_{\text{odd}} = \{1, 3, 5, 7, 9, \ldots\}$$

No matter how far out along the list for W_{even} we go, we'll never find any element that's also in W_{odd}. No matter how far out along the list for W_{odd} we go, we'll never find any element that's also in W_{even}. We won't try to prove this fact now, but your "mathematical sixth sense" should assure you of its truth.

The Venn diagram of Fig. 8-2 illustrates two sets, *J* and *K*, with no elements in common. Imagine *J* as the set of all the points on or inside the circle, and *K* as the set of all the points on or inside the oval. Sets *J* and *K* are disjoint. When we have two disjoint sets, neither of them forms a subset of the other.

Overlapping Sets

When two sets have at least one element in common, we call them *overlapping sets*. Formal texts might call them *nondisjoint sets*. Congruent sets overlap in the strongest possible sense, because they share all of their elements. More often, you'll see cases in which two overlapping sets share some, but not all, of their elements. Following are two sets of numbers that overlap with one element in common:

$$L = \{2, 3, 4, 5, 6\}$$

and

$$M = \{6, 7, 8, 9, 10\}$$

Here's a pair of sets that overlap with several elements in common:

$$P = \{21, 23, 25, 27, 29, 31, 33\}$$

and

$$Q = \{25, 27, 29, 31, 33, 35, 37\}$$

Technically, the following two sets overlap as well:

$$R = \{11, 12, 13, 14, 15, 16, 17, 18, 19\}$$

and

$$S = \{12, 13, 14\}$$

In the last case above, S constitutes a subset of R. In fact, S is a proper subset of R. Now, let's look at a pair of infinite sets that overlap with four elements in common:

$$W_{3-} = \{..., -5, -4, -3, -2, -1, 0, 1, 2, 3\}$$

and

$$W_{0+} = \{0, 1, 2, 3, 4, 5, ...\}$$

The notation W_{3-} (read "W sub three-minus") means the set of integers starting at 3 and decreasing, one by one, without end. The notation W_{0+} (read "W sub zero-plus") means the set of integers starting with 0 and increasing, one by one, without end.

Figure 8-3 shows two sets, T and U, with some elements in common, so they overlap. You can imagine T as the set of all the points on or inside the circle, and U as the set of all the points on or inside the oval. When you have two overlapping sets, one of them can compose a subset of the other, but that doesn't have to be the case. It's clearly not true for T and U in Fig. 8-3.

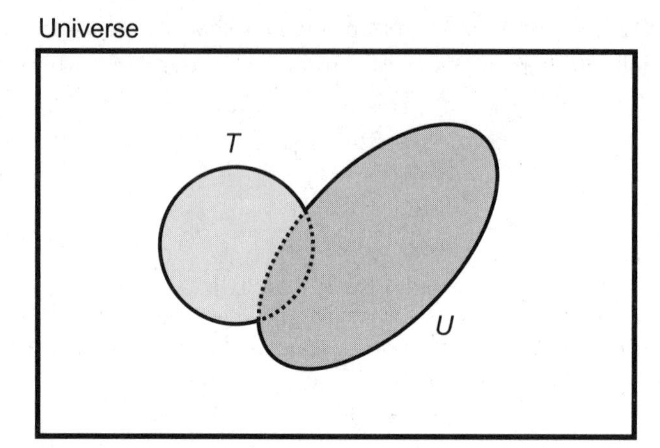

FIGURE 8-3 • Two overlapping sets, T and U. They have *some* elements in common.

PROBLEM 8-4

Is the universal set a subset of itself? Is it a proper subset of itself?

SOLUTION

The universal set (call it U) is a subset of itself. That's a trivial notion, because any set forms a subset of itself. But U is not a proper subset of itself. Remember, we define U as the set of all entities, real or imaginary, past, present, future—everything! If U were a proper subset of itself, then there would exist some entity that did not belong to U. That's impossible because it contradicts the very definition of the universal set!

TIP *The foregoing convoluted argument offers a "massive" example of* reductio ad absurdum, *which you first encountered in Chap. 2.*

PROBLEM 8-5

Provide an example of two sets, both with infinitely many elements, but such that one set forms a proper subset of the other.

SOLUTION

Plenty of set pairs will satisfy this requirement! The set of all even whole numbers (nonnegative integers), W_{even}, is a proper subset of the set of all whole numbers, W. Both of these sets have infinitely many elements.

Amazingly enough, you can pair the elements of both sets off one-to-one, even though one of them is a proper subset of the other! You can get an idea of how this pairing-off works if you divide every element of W_{even} by 2, one at a time, and then write down the first few elements of the resulting set. When you do that, you get

{0/2, 2/2, 4/2, 6/2, 8/2, 10/2, ...}

That's exactly the same as *W*, because when you perform the divisions, you get

{0, 1, 2, 3, 4, 5, ...}

PROBLEM 8-6

List all the subsets of the set {1, 2, 3}. Here's a hint: When you want to denote all the subsets of a small set, start by listing the null set (which is a subset of any set). Then list all of the set's individual elements and enclose them in set braces one by one. Finally list every possible set that contains at least one of those elements.

SOLUTION

First, write down the symbol for the null set. Then isolate the individual elements of {1, 2, 3} and enclose the numerals in braces. Finally assemble and list all the sets you can, using one or more of the elements 1, 2, and 3. You'll get the following list:

∅

{1}

{2}

{3}

{1, 2}

{1, 3}

{2, 3}

{1, 2, 3}

PROBLEM 8-7

List all the subsets of {1, {2, 3}}. Be careful! You must strictly follow the instructions in the hint that goes along with Problem 8-6.

SOLUTION

The set {1, {2, 3}} has only two elements: the number 1 and the set {2, 3}. You can't break {2, 3} down and have it remain an element of the original set {1, {2, 3}}. The list of all possible subsets of {1, {2, 3}} is surprisingly short:

$$\varnothing$$

{1}

{{2, 3}}

{1, {2, 3}}

PROBLEM 8-8

List all the subsets of {1, {2, {3}}}. Be extra careful! The hint given with Problem 8-6 is critical here.

SOLUTION

The set {1, {2, {3}}} has two elements: the number 1 and the set {2, {3}}. You can't break {2, {3}} down and have it remain an element of the original set {1, {2, {3}}}. Therefore, all the possible subsets of {1, {2, {3}}} are as follows:

$$\varnothing$$

{1}

{{2, {3}}}

{1, {2, {3}}}

TIP *When you write down arcane sets such as the foregoing, you must use the same total number of opening braces and closing braces in every set expression. Count up the braces after you've written each item. If there are more or fewer opening braces than closing braces, you've made a mistake somewhere.*

Set Intersection

The *intersection* of two sets comprises *all* the elements, but *only* the elements, that belong to both sets. When you have two sets, say V and W, their intersection is a set that you denote by writing $V \cap W$. The upside-down U-like symbol translates to the word "intersect," so you say "V intersect W."

Intersection of Two Congruent Sets

The intersection of two congruent (identical) sets is the set of all elements in either set. For any two sets X and Y,

$$\text{If } X = Y$$

$$\text{then}$$

$$X \cap Y = X$$

$$\text{and}$$

$$X \cap Y = Y$$

TIP ***Because X and Y are identical, you're actually dealing with only one set in the above situation, not two sets! You can just as well write***

$$X \cap X = X$$

Intersection with the Null Set

The intersection of the null set with any other set always gives you the null set. For any set V,

$$V \cap \varnothing = \varnothing$$

Remember, any element in the intersection of two sets must belong to both of those sets. Nothing can belong to a set that contains no elements! Therefore, nothing can belong to the intersection of the null set with any other set.

Intersection of Two Disjoint Sets

When two sets are disjoint, they have no elements in common, so nothing can belong to them both. The intersection of two disjoint sets always equals the null set. The set size makes no difference. For example, recall the sets of even and odd whole numbers, W_{even} and W_{odd}. They both contain infinitely many elements, but even so,

$$W_{even} \cap W_{odd} = \varnothing$$

Intersection of Two Overlapping Sets

When two sets overlap, their intersection contains at least one element. There's no limit to how many elements the intersection of two sets can have. You have to keep in mind only one requirement: Every element in the intersection set must belong to both of the original sets.

Let's look at the examples of overlapping sets you saw a little while ago, and figure out the intersection sets. First, examine the following two sets:

$$L = \{2, 3, 4, 5, 6\}$$

and

$$M = \{6, 7, 8, 9, 10\}$$

Here, the intersection set contains one element:

$$L \cap M = \{6\}$$

This expression refers to the set containing the number 6, not just the number 6 itself. Now look at the following two sets:

$$P = \{21, 23, 25, 27, 29, 31, 33\}$$

and

$$Q = \{25, 27, 29, 31, 33, 35, 37\}$$

In this case, the intersection set contains five elements:

$$P \cap Q = \{25, 27, 29, 31, 33\}$$

Now check out the following two sets:

$$R = \{11, 12, 13, 14, 15, 16, 17, 18, 19\}$$

and

$$S = \{12, 13, 14\}$$

In this situation, $S \subset R$, so the intersection equals set S. You can write that fact as follows:

$$R \cap S = S = \{12, 13, 14\}$$

How about the set W_{3-} of all positive, negative, or zero whole numbers less than or equal to 3, and the set W_{0+} of all the nonnegative whole numbers? You can write down these two sets as follows:

$$W_{3-} = \{..., -5, -4, -3, -2, -1, 0, 1, 2, 3\}$$

and

$$W_{0+} = \{0, 1, 2, 3, 4, 5, ...\}$$

The intersection set has four elements:

$$W_{3-} \cap W_{0+} = \{0, 1, 2, 3\}$$

The Venn diagram of Fig. 8-4 shows two overlapping sets. Think of the set V as the rectangle and all the points inside. Imagine the set W as the oval and all the points inside. The two regions are hatched diagonally, but in different directions. The intersection set $V \cap W$ shows up as an irregular cross-hatched region.

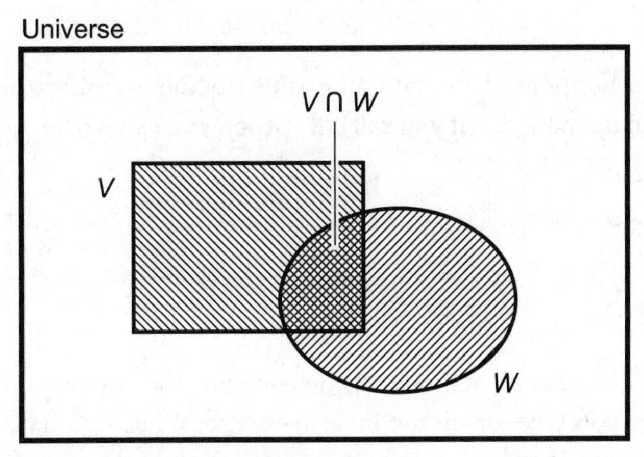

FIGURE 8-4 • Two overlapping sets *V* and *W*. The double-hatched region illustrates their intersection.

PROBLEM 8-9

Find two sets of whole numbers that overlap, with neither set forming a subset of the other, and whose intersection set contains infinitely many elements.

SOLUTION

You can come up with many examples of set pairs like this. Look at the set of all positive whole numbers divisible by 4 without a remainder. (When a quotient has no remainder, that quotient always equals a whole number.) Call this set W_{4d}. Similarly, let W_{6d} be the set of all positive whole numbers divisible by 6 without a remainder. You can list the first few elements of each set as follows:

$$W_{4d} = \{0, 4, 8, 12, 16, 20, 24, 28, 32, 36, \ldots\}$$

and

$$W_{6d} = \{0, 6, 12, 18, 24, 30, 36, 42, 48, \ldots\}$$

Both of the sets W_{4d} and W_{6d} contain infinitely many elements. The two sets overlap because they share certain elements. But neither set forms a subset of the other; they both have some elements all their own. The intersection is the set of all elements divisible by both 4 and 6; call it W_{4d6d}. If you write out both of the above lists up to all values less than or equal to 100 and then identify the elements common to both sets, you'll obtain

$$W_{4d} \cap W_{6d} = W_{4d6d}$$
$$= \{0, 12, 24, 36, 48, 60, 72, 84, 96, \ldots\}$$

This set happens to contain all positive whole numbers divisible by 12 without a remainder. If you call it W_{12d}, then you can write

$$W_{4d} \cap W_{6d} = W_{12d}$$

Set Union

The *union* of two sets contains all of the elements that belong to one set or the other, where you take "or" in the inclusive sense. When you have two sets, say X and Y, their union also forms a set, written $X \cup Y$. You can read the U-like symbol as "union" and say "X union Y."

Union of Two Congruent Sets

When you have two congruent sets, their union equals either set taken alone. For any sets X and Y,

$$\text{If } X = Y$$
$$\text{then}$$
$$X \cup Y = X$$
$$\text{and}$$
$$X \cup Y = Y$$

You're actually dealing with only one set here, so you could just as well write

$$X \cup X = X$$

For the null set you have

$$\varnothing \cup \varnothing = \varnothing$$

TIP ***When you have two congruent sets, their union equals their intersection! This fact might seem trivial right now, but you'll sometimes encounter situations where the congruence of two sets isn't obvious. In cases like that, you can compare the union with the intersection to perform a "congruence test." If the union of two sets corresponds precisely to their intersection, then the sets are congruent.***

Union with the Null Set

The union of the null set with any nonempty set gives you that nonempty set. For any nonempty set X, you can write

$$X \cup \varnothing = X$$

Remember, any element in the union of two sets only has to belong to one of them.

Union of Two Disjoint Sets

When two nonempty sets are disjoint, they have no elements in common, but their union always contains some elements. Consider again the sets of even and odd whole numbers, W_{even} and W_{odd}. Their union equals the set of all the whole numbers:

$$W_{even} \cup W_{odd} = \{0, 1, 2, 3, 4, 5, \ldots\}$$

Union of Two Overlapping Sets

Again, let's look at the same examples of overlapping sets we checked out when we worked with intersection. First, consider the following two sets:

$$L = \{2, 3, 4, 5, 6\}$$

and

$$M = \{6, 7, 8, 9, 10\}$$

The union set is

$$L \cup M = \{2, 3, 4, 5, 6, 7, 8, 9, 10\}$$

The number 6 appears in both sets, but we count it only once in the union. (Any specific element can "belong to a set only once.") Now look at these two sets:

$$P = \{21, 23, 25, 27, 29, 31, 33\}$$

and

$$Q = \{25, 27, 29, 31, 33, 35, 37\}$$

The union set in this case is

$$P \cup Q = \{21, 23, 25, \ldots, 33, 35, 37\}$$

That's all the odd whole numbers between, and including, 21 and 37. We count each individual duplicate element 25 through 33 only once. Now examine the following two sets:

$$R = \{11, 12, 13, 14, 15, 16, 17, 18, 19\}$$

and

$$S = \{12, 13, 14\}$$

In this situation, $S \subset R$, so the union set equals R. We can write that down as

$$\begin{aligned} R \cup S &= R \\ &= \{11, 12, 13, 14, 15, 16, 17, 18, 19\} \end{aligned}$$

We count each individual element 12, 13, and 14 only once. Now consider the sets

$$W_{3-} = \{..., -5, -4, -3, -2, -1, 0, 1, 2, 3\}$$

and

$$W_{0+} = \{0, 1, 2, 3, 4, 5, ...\}$$

Here, the union set consists of all the positive and negative whole numbers, along with 0. Let's write that set as $W_{0\pm}$ (read "W sub zero plus-or-minus"). Then:

$$\begin{aligned} W_{3-} \cup W_{0+} &= W_{0\pm} \\ &= \{..., -5, -4, -3, -2, -1, 0, 1, 2, 3, 4, 5, ...\} \end{aligned}$$

We count each individual element 0, 1, 2, and 3 only once. As you know from basic number theory, $W_{0\pm}$ is the set of *integers*.

Figure 8-5 shows two overlapping sets. Think of X as the rectangle and everything inside it. Imagine Y as the oval and everything inside it. We represent the union of the two sets, $X \cup Y$, as the entire shaded region.

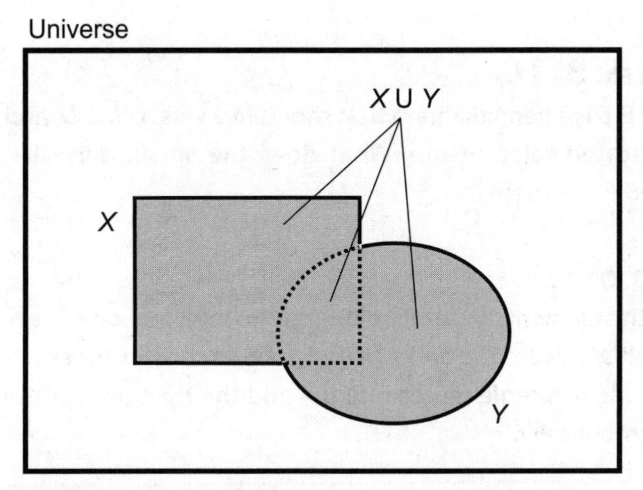

FIGURE 8-5 • Two overlapping sets *X* and *Y*. The entire shaded region illustrates their union.

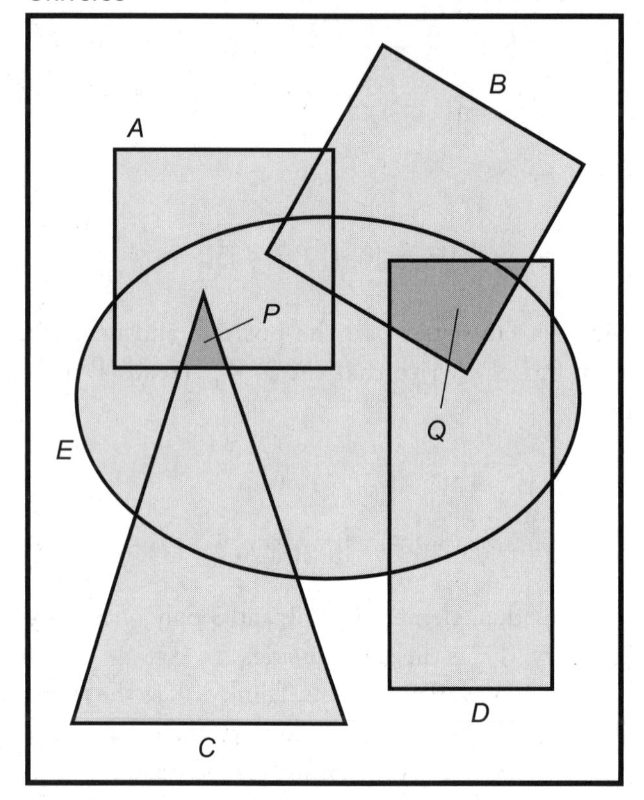

FIGURE 8-6 • Illustration for Problems and Solutions 8-10 through 8-12.

PROBLEM 8-10

Figure 8-6 is a Venn diagram that shows five sets *A, B, C, D,* and *E*, in a rather complicated relationship. What does the small, dark-shaded triangle marked *P* represent?

SOLUTION

All of the elements in set *P* belong to the intersection of sets *A* and *C*. You know this because the dark-shaded triangle constitutes exactly that region where the rectangle representing *A* and the triangle representing *C* overlap. Symbolically,

$$P = A \cap C$$

PROBLEM 8-11

What set does the dark-shaded, irregular, four-sided figure *Q* represent in Fig. 8-6?

SOLUTION

The points in region *Q* are fully shared by the points in *B* and *D*. Therefore, region *Q* represents the intersection of sets *B* and *D*, which you can write as

$$Q = B \cap D$$

PROBLEM 8-12

If you consider all the possible intersections of two sets in Fig. 8-6, which of those intersection sets are empty?

SOLUTION

Whenever two regions share no area at all, the sets they represent are disjoint, so their intersection set is empty. In Fig. 8-6, the only pairs of regions that don't overlap are *A* and *D*, *B* and *C*, and *C* and *D*. Therefore, the null-set intersection pairs in this scenario are as follows:

$$A \cap D = \varnothing$$

$$B \cap C = \varnothing$$

$$C \cap D = \varnothing$$

PROBLEM 8-13

What is the intersection of the following two sets? What is their union?

$$A = \{1, 1/2, 1/3, 1/4, 1/5, 1/6, \ldots\}$$

and

$$G = \{1, 1/2, 1/4, 1/8, 1/16, 1/32, \ldots\}$$

In set *A*, the denominator of the fraction increases by 1 as you go down the list. In set *G*, the denominator doubles as you go down the list. All the numerators in both sets are equal to 1.

SOLUTION

Notice that set *G* contains all the elements, but only those elements, that belong to both sets. Therefore, the intersection of sets *A* and *G* equals set *G*. You symbolize this fact by writing

$$A \cap G = G$$

If you start with set *A* and then add in all the elements of *G*, you get the same set *A* (with certain elements listed twice, but they can count only once). Therefore, set *A* contains precisely those elements that belong to one set or the other. The union of the two sets equals set *A*. You symbolize this fact as

$$A \cup G = A$$

PROBLEM 8-14

Can you find two sets of whole numbers, with one of the sets infinite, but such that their union contains only a finite number of elements?

SOLUTION

No, you can't. Any element in the union of two sets must belong to one or both sets. If a set has infinitely many elements, then the union of that set with any other set—even the null set—must have infinitely many elements.

QUIZ

You may refer to the text in this chapter while taking this quiz. A good score is at least 8 correct. Answers are in the back of the book.

1. **Which of the following statements holds true under all circumstances?**
   A. If *A* is a proper subset of *B*, then *A* is a subset of *B*.
   B. If *A* is a subset of *B*, then *A* is a proper subset of *B*.
   C. If *A* and *B* are identical sets, then *A* is a proper subset of *B*.
   D. Sets *A* and *B* are coincident if and only if *A* is a subset of *B*.

2. **Suppose that you have two nonempty sets *P* and *Q*. Which of the following statements can't hold true under any circumstances?**
   A. $P \cap Q$ is infinite.
   B. $P \cup Q$ is infinite.
   C. $P \cap Q$ is empty.
   D. $P \cup Q$ is empty.

3. **Figure 8-7 illustrates two nonempty sets *E* and *P*. The entire shaded region represents the set of all elements that belong to**
   A. set *E* and set *P*.
   B. set *E* if and only if set *P*.
   C. set *E* or set *P*.
   D. set *E* with set *P*.

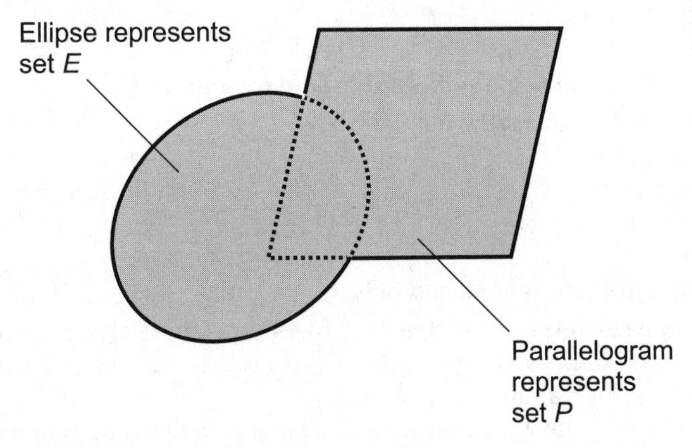

FIGURE 8-7 • Illustration for Quiz Question 3.

4. **How many elements does the set {1, 2, 3, 4, 3, 2, 1} have?**
   A. Three.
   B. Four.
   C. Seven.
   D. We can't say because the list is redundant.

5. **What is the union of {1, 2, 3, 4, 3, 2, 1} with the set of all strictly negative integers?**
   A. $\varnothing$
   B. {..., −4, −3, −2, −1}
   C. {..., −4, −3, −2, −1, 1, 2, 3, 4}
   D. {1, 2, 3, 4, ...}

6. **Consider two sets G and H, such that $G \cap H \neq \varnothing$. In this situation,**
   A. G and H overlap.
   B. G and H are disjoint.
   C. G and H share a single element.
   D. G and H are congruent.

7. **Suppose that you encounter two nonempty sets P and Q. The set $P \cap Q$ contains exactly those elements belonging to**
   A. both P and Q.
   B. P or Q, or both.
   C. either P or Q, but not both.
   D. the universal set.

8. **What is the intersection of {1, 2, 3, 4, 3, 2, 1} with the set of all strictly negative integers?**
   A. $\varnothing$
   B. {..., −4, −3, −2, −1}
   C. {..., −4, −3, −2, −1, 1, 2, 3, 4}
   D. {1, 2, 3, 4}

9. **Which of the following sets is a subset of the null set?**
   A. The set containing the null set.
   B. No such set exists.
   C. The null set.
   D. The set of all sets.

10. **Two sets are congruent if and only if**
    A. they have the same number of elements, and those elements can be paired off in a one-to-one correspondence.
    B. neither set is a subset of the other.
    C. they both contain infinitely many elements that can be paired off in a one-to-one correspondence.
    D. their elements are identical and can be paired off in a one-to-one correspondence.

chapter 9

The Logic of Machines

We call an electronic signal *digital* when it can attain a limited number of well-defined states. Digital signals contrast with *analog* signals, which can vary over a continuous range, and which can therefore attain a theoretically infinite number of possible instantaneous states. Figure 9-1 shows an example of an analog signal (at A) and an example of a digital signal (at B). In the analog case, the signal strength, also called the *amplitude*, varies "smoothly" as time passes. In the digital case shown here, the amplitude can attain only two states.

CHAPTER OBJECTIVES

In this chapter, you will

- Compare decimal, binary, octal, and hexadecimal numeration systems.
- Discover how logic gates work.
- See clocks, counters, and black boxes in action.
- Compare storage and speed units.
- Learn the difference between analog and digital data.
- See how circuits can convert between different data formats.

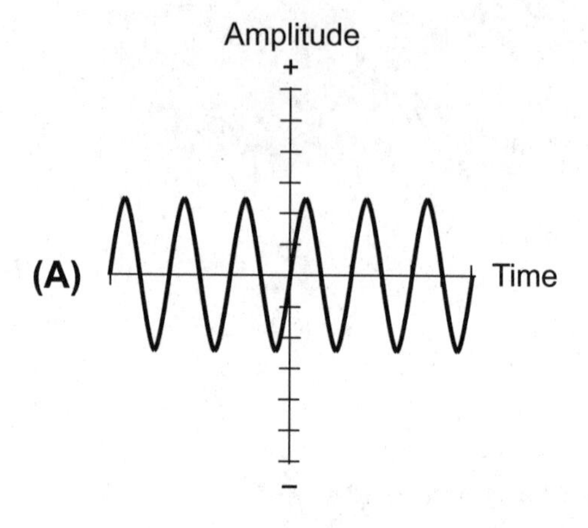

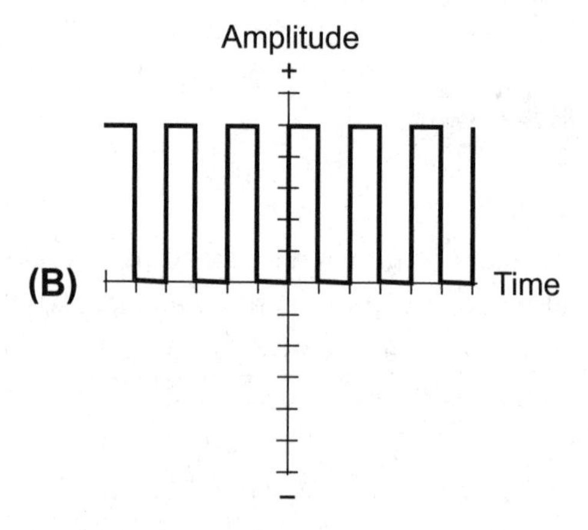

FIGURE 9-1 • An example of an analog electrical signal (A) that can theoretically have infinitely many states, and an example of a digital signal (B) with only two possible states.

Numeration Systems

In everyday life, most of us deal with the *decimal number system*, which makes use of digits from the set

$$D = \{0, 1, 2, 3, 4, 5, 6, 7, 8, 9\}$$

Machines such as computers and communications devices more often use numeration schemes that have some power of 2 digits, such as 2 (2^1), 4 (2^2), 8 (2^3), 16 (2^4), 32 (2^5), 64 (2^6), and so on.

Decimal

The *decimal number system* is also called *base 10* or *radix 10*. When we express nonnegative integers in this system, we multiply the right-most digit by 10^0, or 1. We multiply the next digit to the left by 10^1, or 10. The power of 10 increases as we move further to the left. Once we've done the digit multiplication, we add up all of the resulting values. For example:

$$\begin{array}{c} 8 \times 10^0 \\ 5 \times 10^1 \\ 0 \times 10^2 \\ 2 \times 10^3 \\ 6 \times 10^4 \\ 8 \times 10^5 \\ \hline 862{,}058 \end{array}$$

Binary

The *binary number system* provides us with a method of expressing numbers using only the digits 0 and 1. We'll sometimes hear this system called *base 2* or *radix 2*. When we express nonnegative integers in binary notation, we multiply the rightmost digit by 2^0, or 1. The next digit to the left is multiplied by 2^1, or 2. The power of 2 increases as we continue to the left, so we get a "fours" digit, then an "eights" digit, then a "16s" digit, and so on. For example, consider the decimal number 94. In the binary system, we would write this quantity as 1011110. It breaks down in columnar-sum form as follows:

$$\begin{array}{c} 0 \times 2^0 \\ 1 \times 2^1 \\ 1 \times 2^2 \\ 1 \times 2^3 \\ 1 \times 2^4 \\ 0 \times 2^5 \\ 1 \times 2^6 \\ \hline 94 \end{array}$$

TIP ***When we work with a computer or calculator, we input a decimal number to the machine, which converts the number into binary form. The computer or calculator performs all of its arithmetic operations using only the digits 0 and 1. When the computation or calculation process is complete, the machine converts the result back into decimal form for display. In a digital communications or data-storage system, binary numbers can represent alphanumeric characters, shades of color, degrees of temperature, compass directions, acoustic frequencies, and other variable quantities.***

Octal

Another scheme, sometimes used in computer programming, goes by the name *octal number system* because it has eight symbols (according to our way of thinking), or 2^3. Every digit constitutes an element of the set

$$O = \{0, 1, 2, 3, 4, 5, 6, 7\}$$

Some people call this system *base 8* or *radix 8*. When we express nonnegative integers in octal notation, we multiply the rightmost digit by 8^0, or 1. The next digit to the left is multiplied by 8^1, or 8. The power of 8 increases as we continue to the left, so we get a "64s" digit, then a "512s" digit, and so on. For example, we render the decimal quantity 3085 in octal form as 6015. We can break it down into the following sum:

$$\begin{array}{c} 5 \times 8^0 \\ 1 \times 8^1 \\ 0 \times 8^2 \\ 6 \times 8^3 \\ \hline 3085 \end{array}$$

Hexadecimal

Another system used in computer work is the *hexadecimal number system*. It has 16 (2^4) symbols: the usual 0 through 9 plus six more, represented by the uppercase English letters A through F. We therefore get the digit set

$$H = \{0, 1, 2, 3, 4, 5, 6, 7, 8, 9, A, B, C, D, E, F\}$$

This system is sometimes called *base 16* or *radix 16*. All of the hexadecimal digits 0 through 9 represent the same values as their decimal counterparts. However, we have the following additional digits:

- Hexadecimal A equals decimal 10
- Hexadecimal B equals decimal 11
- Hexadecimal C equals decimal 12
- Hexadecimal D equals decimal 13
- Hexadecimal E equals decimal 14
- Hexadecimal F equals decimal 15

When we express nonnegative integers in hexadecimal notation, we multiply the rightmost digit by 16^0, or 1. We multiply the next digit to the left by 16^1, or 16. The power of 16 increases as we continue to the left, so we get a "256s" digit, then a "4096s" digit, and so on. For example, we write the decimal quantity 35,898 in hexadecimal form as 8C3A. Remembering that C = 12 and A = 10, we can break the hexadecimal number down into the following sum:

$$\begin{array}{c} A \times 16^0 \\ 3 \times 16^1 \\ C \times 16^2 \\ 8 \times 16^3 \\ \hline 35{,}898 \end{array}$$

Table 9-1 lists the binary, octal, and hexadecimal equivalents for the decimal numbers 0 through 64.

TABLE 9-1 Binary, octal, and hexadecimal equivalents for decimal numbers 0 through 64.

Decimal	Binary	Octal	Hexadecimal
0	0	0	0
1	1	1	1
2	10	2	2
3	11	3	3
4	100	4	4

TABLE 9-1 Binary, octal, and hexadecimal equivalents for decimal numbers 0 through 64. (*Continued*)

Decimal	Binary	Octal	Hexadecimal
5	101	5	5
6	110	6	6
7	111	7	7
8	1000	10	8
9	1001	11	9
10	1010	12	A
11	1011	13	B
12	1100	14	C
13	1101	15	D
14	1110	16	E
15	1111	17	F
16	10000	20	10
17	10001	21	11
18	10010	22	12
19	10011	23	13
20	10100	24	14
21	10101	25	15
22	10110	26	16
23	10111	27	17
24	11000	30	18
25	11001	31	19
26	11010	32	1A
27	11011	33	1B
28	11100	34	1C
29	11101	35	1D
30	11110	36	1E
31	11111	37	1F
32	100000	40	20
33	100001	41	21
34	100010	42	22
35	100011	43	23
36	100100	44	24

TABLE 9-1 Binary, octal, and hexadecimal equivalents for decimal numbers 0 through 64. (*Continued*)

Decimal	Binary	Octal	Hexadecimal
37	100101	45	25
38	100110	46	26
39	100111	47	27
40	101000	50	28
41	101001	51	29
42	101010	52	2A
43	101011	53	2B
44	101100	54	2C
45	101101	55	2D
46	101110	56	2E
47	101111	57	2F
48	110000	60	30
49	110001	61	31
50	110010	62	32
51	110011	63	33
52	110100	64	34
53	110101	65	35
54	110110	66	36
55	110111	67	37
56	111000	70	38
57	111001	71	39
58	111010	72	3A
59	111011	73	3B
60	111100	74	3C
61	111101	75	3D
62	111110	76	3E
63	111111	77	3F
64	1000000	100	40

PROBLEM 9-1

Express the binary number 10011011 in decimal form.

SOLUTION

We can add the digits up by going from left to right, or from right to left. It doesn't matter which way we go, but most people find it easier to keep track of the progression by starting at the "ones" place and working toward the left from there. The digits add up as follows:

$$\begin{array}{c} 1 \times 2^0 \\ 1 \times 2^1 \\ 0 \times 2^2 \\ 1 \times 2^3 \\ 1 \times 2^4 \\ 0 \times 2^5 \\ 0 \times 2^6 \\ 1 \times 2^7 \\ \hline 155 \end{array}$$

PROBLEM 9-2

Express the decimal number 1,000,000 in hexadecimal form. Warning: This problem involves some tedious arithmetic!

SOLUTION

The values of the digits in a whole (nonfractional) hexadecimal number, proceeding from right to left, constitute ascending nonnegative integer powers of 16. Therefore, a whole hexadecimal number n_{16} has the form

$$n_{16} = ... + (f \times 16^5) + (e \times 16^4) + (d \times 16^3) + (c \times 16^2) + (b \times 16^1) + (a \times 16^0)$$

where *a*, *b*, *c*, *d*, *e*, *f*, ... represent single-digit hexadecimal numbers from the set

$$H = \{0, 1, 2, 3, 4, 5, 6, 7, 8, 9, A, B, C, D, E, F\}$$

Let's begin by finding the largest power of 16 that's less than or equal to 1,000,000. This is $16^4 = 65{,}536$. Then, we divide 1,000,000 by 65,536, obtaining 15 plus a remainder. We represent the decimal 15 as the hexadecimal F. Now we know that the hexadecimal expression of the decimal number 1,000,000 has the form

$$(\mathrm{F} \times 16^4) + (d \times 16^3) + (c \times 16^2) + (b \times 16^1) + a$$
$$= \mathrm{F}dcba$$

To find the value of *d*, we note that

$$15 \times 16^4 = 983{,}040$$

This quantity is 16,960 smaller than 1,000,000, so we must find the hexadecimal equivalent of decimal 16,960 and add it to hexadecimal F0000. The largest power of 16 that's less than or equal to 16,960 is 16^3, or 4096. We divide 16,960 by 4096 to obtain 4 plus a remainder. Now we know that $d = 4$ in the above expression, so the decimal 1,000,000 has the hexadecimal form

$$(\mathrm{F} \times 16^4) + (4 \times 16^3) + (c \times 16^2) + (b \times 16^1) + a$$
$$= \mathrm{F4}cba$$

To find the value of *c*, we note that

$$(\mathrm{F} \times 16^4) + (4 \times 16^3) = 983{,}040 + 16{,}384$$
$$= 999{,}424$$

This quantity is 576 smaller than 1,000,000, so we must find the hexadecimal equivalent of decimal 576 and add it to hexadecimal F4000. The largest power of 16 that's less than or equal to 576 is 16^2, or 256. We divide 576 by 256 to get 2 plus a remainder. Now we know that $c = 2$ in the above expression, so the decimal 1,000,000 is equivalent to the hexadecimal

$$(\mathrm{F} \times 16^4) + (4 \times 16^3) + (2 \times 16^2) + (b \times 16^1) + a$$
$$= \mathrm{F42}ba$$

To find the value of *b*, we note that

$$(\mathrm{F} \times 16^4) + (4 \times 16^3) + (2 \times 16^2) = 983{,}040 + 16{,}384 + 512$$
$$= 999{,}936$$

This quantity is 64 smaller than 1,000,000, so we must find the hexadecimal equivalent of decimal 64 and add it to hexadecimal F4200. The largest power of 16 that's less than or equal to 64 is 16^1, or 16. We divide 64 by 16, obtaining 4 without any remainder. Now we know that $b = 4$ in the above expression, so the decimal 1,000,000 is equivalent to the hexadecimal

$$(F \times 16^4) + (4 \times 16^3) + (2 \times 16^2) + (4 \times 16^1) + a$$
$$= F424a$$

When we found *b*, we were left with no remainder. Therefore, all the digits to the right of *b* (in this case, that means only the digit *a*) must equal 0. We have dragged ourselves through a tedious process, to be sure; but we've finally determined that the decimal number 1,000,000 equals the hexadecimal number F4240. To check our work, we can do our arithmetic the other way, converting the hexadecimal number to decimal form, multiplying out the digits from right to left. That's a lot easier:

$$0 \times 16^0$$
$$4 \times 16^1$$
$$2 \times 16^2$$
$$4 \times 16^3$$
$$15 \times 16^4$$
$$\overline{1{,}000{,}000}$$

Digital Circuits

All binary (two-state) digital devices and systems employ high-speed electronic switches that perform boolean operations. Engineers call these switches *logic gates*. Even the most advanced supercomputers consist of logic gates when we break them down to the elemental level.

Positive versus Negative Logic

We might suppose that in electronic logic, the binary digit 1 should stand for "truth" while the binary digit 0 should stand for "falsity." In so-called *positive logic*, things work that way. A circuit represents the binary digit 1 with an *electrical potential* of approximately +5 *volts* (called the *high state* or simply *high*), while the binary digit 0 appears as little or no voltage (called the *low state* or simply *low*).

Some electronic systems employ *negative logic*, in which little or no voltage (the low state) represents logic 1, while +5 volts (the high state) represents logic 0. In another form of negative logic, the digit 1 appears as a negative voltage (such as –5 volts, constituting the low state) and the digit 0 appears as little or no voltage (the high state in this case, because it has the more positive voltage).

TIP ***In practice, it doesn't matter what voltage represents which logic state. An electronic logic circuit will work perfectly well as long as the two voltages differ significantly, and as long as they always represent the same logic states as time passes. For the remainder of this discussion, let's stick to positive logic, where low means logic 0 and high means logic 1.***

Logic Gates

All digital electronic devices employ switches that perform specific logical operations. These switches, called *logic gates*, can have anywhere from one to several inputs and (usually) a single output. Six basic types of logic gates exist, as follows:

- A *logical inverter*, also called a *NOT gate*, has one input and one output. It simply reverses, or inverts, the state of the input. If the input equals 1, then the output equals 0. If the input equals 0, then the output equals 1.
- An *OR gate* can have two or more inputs (although it usually has only two). If both, or all, of the inputs equal 0, then the output equals 0. If any of the inputs equal 1, then the output equals 1. It only takes one "true" input to make the output of the OR gate "true." The gate therefore performs an inclusive-OR operation.
- An *AND gate* can have two or more inputs (although it usually has only two). If both, or all, of the inputs equal 1, then the output equals 1. If any of the inputs equal 0, then the output equals 0.
- An OR gate can be followed by a NOT gate. This combination gives us a *NOT-OR* gate, more often called a *NOR gate*. If both, or all, of the inputs equal 0, then the output equals 1. If any of the inputs equal 1, then the output equals 0.
- An AND gate can be followed by a NOT gate. This combination gives us a *NOT-AND* gate, more often called a *NAND gate*. If both, or all, of the inputs equal 1, then the output equals 0. If any of the inputs equal 0, then the output equals 1.

- An *exclusive OR gate,* also called an *XOR gate,* has two inputs and one output. If the two inputs have the same state (either both 1 or both 0), then the output equals 0. If the two inputs have different states, then the output equals 1.

Table 9-2 summarizes the functions of these logic gates, assuming a single input for the NOT gate and two inputs for the others. Figure 9-2 illustrates the schematic symbols that engineers and technicians use to represent these gates in circuit diagrams.

TABLE 9-2 Logic gates and their characteristics.

Gate type	Number of inputs	Remarks
NOT	1	Changes state of input
OR	2 or more	Output high if any inputs are high Output low if all inputs are low
AND	2 or more	Output low if any inputs are low Output high if all inputs are high
NOR	2 or more	Output low if any inputs are high Output high if all inputs are low
NAND	2 or more	Output high if any inputs are low Output low if all inputs are high
XOR	2	Output high if inputs differ Output low if inputs are the same

Black Boxes

Electronics and computer engineers can combine digital logic gates to create massive, almost unbelievably complicated systems having millions or billions of inputs and outputs. No matter how complicated the system might get, the outputs always constitute specific *logical functions* of the inputs. Engineers sometimes call a complex combination of logic gates a *black box.*

If an engineer knows the functions of all the logic gates inside a black box, and also knows exactly how the gates are interconnected, then the engineer can define the boolean function that the box performs. Conversely, if an engineer needs a black box to execute a specific logical function, then the engineer can use boolean algebra to break the function down into the six basic operations NOT, OR, AND, NOR, NAND, and XOR, and assemble the gates accordingly.

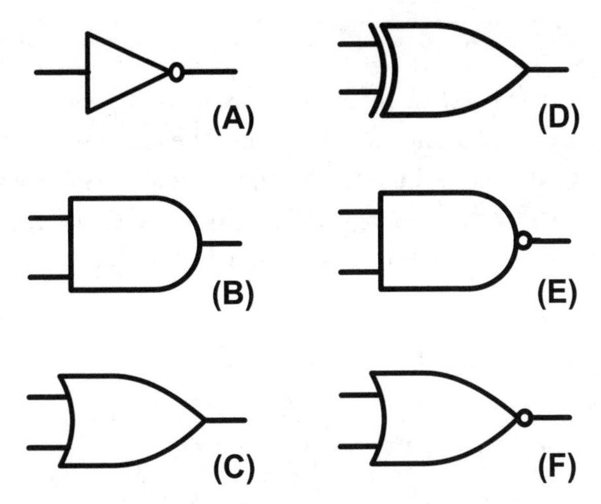

FIGURE 9-2 • Schematic symbols for an inverter or NOT gate (A), an AND gate (B), an OR gate (C), an XOR gate (D), a NAND gate (E), and a NOR gate (F). Inputs appear on the left-hand sides of the symbols. Outputs appear on the right-hand sides.

Forms of Binary Data

In communications systems, binary data does not exhibit as much sensitivity to *noise* and *interference* as analog data does. For this reason, binary systems usually work better than analog systems, producing fewer errors and allowing communication under more adverse circumstances. Since the first digital transmissions took place more than 100 years ago, engineers have invented numerous forms, or *modes*, of binary communication. Following are three historically well-known *binary signaling* modes.

- *Morse code* is the oldest two-state means of sending and receiving messages. The logic states are known as *mark* (key-closed or on) and *space* (key-open or off). Morse code is largely obsolete, but some amateur radio operators still use it in their hobby activities, just for fun!
- *Baudot*, also called the *Murray code*, is a five-unit digital code not widely used by today's digital equipment, except in a few antiquated *teleprinter* systems. There exist 2^5, or 32, possible representations.
- The *American National Standard Code for Information Interchange* (ASCII) is a seven-unit code for the transmission of text and simple computer programs. There exist 2^7, or 128, possible representations.

Clocks

In electronics, the term *clock* refers to a circuit that generates pulses at high speed and at precise, constant time intervals. The clock sets the tempo for the operation of digital devices. In a computer, the clock acts like a metronome for the *microprocessor*. We express or measure clock speeds as frequencies in *hertz* (Hz). One hertz equals one pulse per second. Higher-frequency units work out as follows:

- A *kilohertz* (kHz) equals 1000 Hz
- A *megahertz* (MHz) equals 1,000,000 Hz
- A *gigahertz* (GHz) equals 1,000,000,000 Hz
- A *terahertz* (THz) equals 1,000,000,000,000 Hz

In positive logic, a clock generates brief high pulses at regular intervals. The normal state is low.

Counters

A *counter* constitutes an electronic circuit that does exactly what its name suggests: It literally counts digital pulses. Each time the counter receives a high pulse, the binary number in its *memory* increases by 1. A *frequency counter* can measure the frequency of an alternating-current (AC) wave or signal by tallying up the cycles over a precisely known period of time. The circuit consists of a *gate*, which begins and ends each counting cycle at defined intervals. (Don't confuse this type of gate with the logic gates described a few moments ago!) The counter's accuracy depends on the *gate time*, or how long the gate remains open to accept pulses for counting. As we increase the gate time in a frequency counter, accuracy improves.

TIP ***Although the counter tallies up the pulses as binary numbers, the readout usually appears in base-10 digital numerals for user convenience.***

PROBLEM 9-3

Suppose that we place a NOT gate in *series* (i.e., in cascade) with each of the two inputs of an AND gate. Under what conditions does the output of the resulting black box appear high?

SOLUTION

Figure 9-3 illustrates the arrangement of logic gates as a *schematic diagram*. The inputs appear at points *X* and *Y*. We define two intermediate circuit points *P* and *Q*, and we call the output point *R*. At *P*, we have the condition "NOT *X*." At *Q*, we have the condition "NOT *Y*." At *R*, we have the condition

(NOT *X*) AND (NOT *Y*)

When we compile all of the possible logic states at each of these points, we get Table 9-3, which tells us that the output of the black box is high (logic 1) if and only if both input states are low (logic 0).

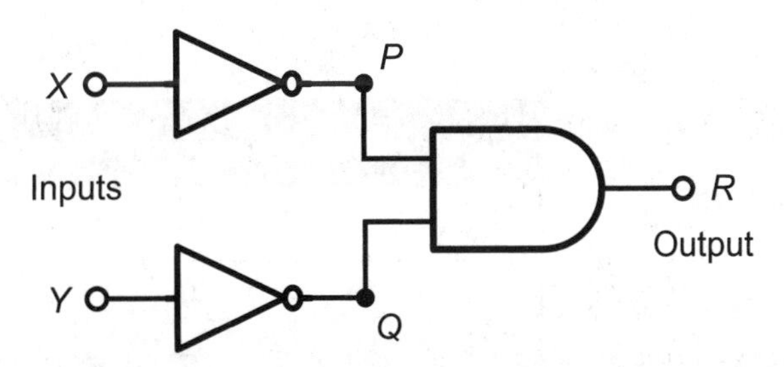

FIGURE 9-3 • Illustration for the solution to Problem 9-3.

TABLE 9-3 Solution to Problem 9-3.

X	*Y*	*P*	*Q*	*R*
0	0	1	1	1
0	1	1	0	0
1	0	0	1	0
1	1	0	0	0

PROBLEM 9-4

Suppose that an XOR gate is followed by a NOT gate. Under what conditions does the output of the resulting black box appear low?

SOLUTION

Figure 9-4 shows the sequence of logic gates for this scenario. As before, we call the inputs *X* and *Y*. At point *P*, we have the condition "*X* NOR *Y*," which is high if *X* and *Y* have opposite logic states, and low if *X* and *Y* have the same logic state. At point *Q*, we have the output condition, which reverses the logic state at point *P*. Table 9-4 depicts all possible logic states at points *X*, *Y*, *P*, and *Q*. This table tells us that the output of the black box is low if and only if the input states oppose each other.

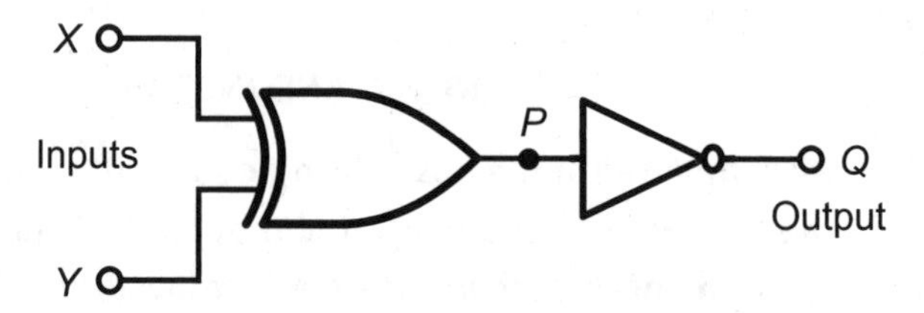

FIGURE 9-4 • Illustration for the solution to Problem 9-4.

TABLE 9-4 Solution to Problem 9-4.

X	*Y*	*P*	*Q*
0	0	0	1
0	1	1	0
1	0	1	0
1	1	0	1

PROBLEM 9-5

Suppose that we place a NOT gate in series with each of the inputs of an XOR gate. Under what conditions will the output of the resulting black box appear high?

SOLUTION

Figure 9-5 shows the arrangement of logic gates in this situation. At point *P*, we have the condition "NOT *X*." At point *Q*, we have the condition "NOT *Y*." At point *R*, we have

(NOT *X*) XOR (NOT *Y*)

Table 9-5 is a compilation of all possible states. We can see that the output of the black box is high if and only if the input states are opposite.

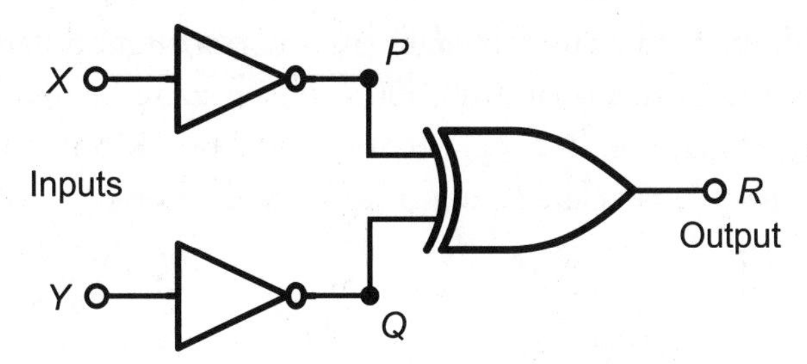

FIGURE 9-5 • Illustration for the solution to Problem 9-5.

TABLE 9-5 Solution to Problem 9-5.

X	*Y*	*P*	*Q*	*R*
0	0	1	1	0
0	1	1	0	1
1	0	0	1	1
1	1	0	0	0

Digital Signals

The use of binary data provides excellent communications accuracy, efficiency, and reliability. If we need to have more than two logic states (a condition known as *multilevel signaling*), then we can represent all the levels as sequences of binary digits. A sequence of three binary digits, for example, can represent 2^3, or eight, levels. A sequence of four binary digits can represent 2^4, or 16, levels. Communications engineers and technophiles compress the term *binary digit* into the single word *bit*.

Bits

We can always represent a digital bit as either logic 0 or logic 1. A group of eight bits constitutes a unit called an *octet*. In many systems, an octet corresponds to a *byte*. We can express in large units according to powers of 10, as follows:

- A *kilobit* (kb) equals 1000 bits
- A *megabit* (Mb) equals 1,000,000 bits
- A *gigabit* (Gb) equals 1,000,000,000 bits
- A *terabit* (Tb) equals 1,000,000,000,000 bits

If you hear about a computer *modem* (modulator/demodulator) that operates at 50 Mbps, it means 50,000,000 *bits per second* (bps). Bits, kilobits, megabits, gigabits, and terabits per second (bps, kbps, Mbps, Gbps, and Tbps) express *data speeds*, also called *data transfer rates*, in digital communications networks.

Bytes

Data that resides in computer storage or memory (as opposed to "flowing" from point to point in a network) is usually quantified as follows:

- A *kilobyte* (KB) equals 2^{10} or 1024 bytes
- A *megabyte* (MB) equals 2^{20} or 1,048,576 bytes
- A *gigabyte* (GB) equals 2^{30} or 1,073,741,824 bytes
- A *terabyte* (TB) equals 2^{40} or 1,099,511,627,776 bytes

TIP ***Note that we abbreviate bits as a lowercase b, but we abbreviate bytes as an uppercase B. We represent 10^3 (the decimal prefix multiplier* kilo-*) as a lowercase k, but 2^{10} (the binary prefix multiplier* kilo-*) gets an uppercase K. The prefix multipliers* mega-, giga-, *and* tera- *are abbreviated as uppercase letters M, G, and T, respectively, whether we're working with "bunches of bits" (powers of 10) or "bunches of bytes" (powers of 2).***

Larger units for* static data *(stored as opposed to moving) have become common in recent years, as computer memory and storage media continue to expand in capacity. Therefore, you should expect to encounter units such as the following more and more often in the future:

- ***A* petabyte *(PB) equals 2^{50} bytes, or 1024 TB***
- ***An* exabyte *(EB) equals 2^{60} bytes, or 1024 PB***

Baud

The term *baud* refers to the number of times per second that a signal changes its logic state. The units of bps (bits per second) and baud are *not* equivalent, although people often think or speak of them that way. You won't encounter this term very often.

TABLE 9-6 Lengths of time required to send text data at various speeds. Abbreviations are as follows: kbps = kilobits per second (units of 1000 bits per second), Mbps = megabits per second (units of 1,000,000 bits per second), s = seconds, ms = milliseconds (units of 0.001 second), and μs = microseconds (units of 0.000001 second).

A			
Speed, kbps	**Time for one page**	**Time for 10 pages**	**Time for 100 pages**
28.8	0.38 s	3.8 s	38 s
38.4	280 ms	2.8 s	28 s
57.6	190 ms	1.9 s	19 s
100	110 ms	1.1 ms	11 s
250	44 ms	440 ms	4.4 s
500	22 ms	220 ms	2.2 s

B			
Speed, Mbps	**Time for one page**	**Time for 10 pages**	**Time for 100 pages**
1.00	11 ms	110 ms	1.1 s
2.50	4.4 ms	44 ms	440 ms
10.0	1.1 ms	11 ms	110 ms
100	110 μs	1.1 ms	11 ms

Examples of Data Speed

In a computer network, each computer or terminal has a modem connecting it to the communications medium (cable, optical fiber, wireless link, telephone line, or whatever). When two machines exchange digital data, the slower modem determines the maximum possible data speed. Table 9-6 shows some examples of data speeds and the approximate time periods required to send 1, 10, and 100 pages of double-spaced, typewritten text at each speed.

Analog-to-Digital Conversion

We can convert any analog signal into a string of pulses whose amplitudes can attain a limited number of digital states. We call this process *analog-to-digital* (A/D) *conversion*. In a computer or communications system, we can use an *A/D converter* to change voices, pictures, music, or test-instrument readings into signals that a digital machine can "understand." In a high-fidelity audio recording system, we can convert music from analog to digital form for storage on media such as *compact discs* (CDs) or *digital versatile discs* (DVDs). Digital signals suffer much less than analog signals from the adverse effects of noise and interference. Also, we can get many more digital signals than analog signals into any given circuit, onto any given communications line, or onto any given storage medium.

An A/D converter works by "checking" the level of an analog signal every so-often, and then rounding it off to the nearest standard digital level, a process known as *sampling*. An analog signal resembles a smooth hillside, while the equivalent digital signal resembles a terraced hillside with each terrace representing one sample. Some A/D converters have only a few levels, or states, in their outputs; others have many. We call the number of possible digital states the *sampling resolution*. Some A/D converters take many samples every second, while others take only a few. The number of samples taken per second constitutes the *sampling rate*.

In most A/D converters, the sampling resolution equals a positive-integer power of 2. A sampling resolution of eight levels (Fig. 9-6) is sufficient for voice communications. A resolution of 16 levels is good enough for music transmission or recording of fair quality. For video signals, such as we might find in high-speed *animated computer graphics* or *virtual reality* (VR), we need a higher sampling resolution.

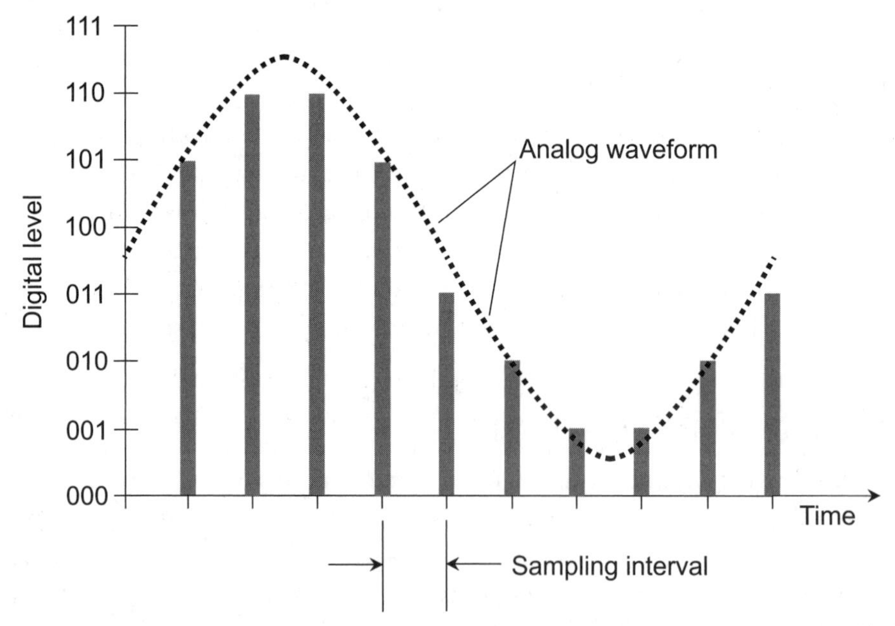

FIGURE 9-6 • An analog waveform (dashed curve) and an eight-level digital representation of that waveform (vertical bars).

The precision with which we can digitize an analog signal depends on the sampling rate. In general, the sampling rate must equal at least twice the highest data frequency if we expect to get a reasonable digital facsimile of a given analog signal. For a voice signal with components as high as 3 kHz, the minimum sampling rate is 6 kHz, representing one sample every 167 microseconds (μs), where 1 μs = 0.000001 second (s). In industrial-level *computer-aided design* (CAD), *gaming, high-definition television* (HDTV), or *virtual reality* (VR), we will need a sampling rate on the order of several megahertz.

Digital-to-Analog Conversion

Digital-to-analog (D/A) *conversion* does the opposite of A/D conversion. The D/A process converts a digital signal to an equivalent analog signal at the receiving end (or *destination*) in a communications circuit where A/D conversion is done at the transmitting end (or *source*). We can use D/A conversion to reproduce digitized sound recordings, such as those found on a CD, DVD, computer *flash drive*, or computer *hard drive*.

In recording and reproduction, we can copy a digital selection repeatedly, and the quality does not diminish. With older analog recording and reproduction equipment, the fourth or fifth "generation" always ended up sounding or looking bad, regardless of the quality of the equipment used. Of course, no data transmission, storage, or retrieval system works perfectly. Errors occasionally creep into digital systems. A circuit will sometimes mistake a logic 1 for a logic 0 or vice versa. Such errors accumulate over time, and as generation after generation of copies are made. But in general, digital systems represent a huge technological improvement over analog systems, because digital errors occur far less often, and have less effect overall, than analog imperfections.

The methodology for D/A conversion depends on whether the signal is purely binary (consisting of only two logic states) or multilevel, with a number of states corresponding to some integer power of 2 (such as 4, 8, 16, or 32). Generally, D/A conversion is done by a microprocessor that undoes what an A/D converter originally did. We can convert a multilevel digital signal to its analog equivalent by "smoothing out" the pulses. (Imagine the train of pulses in Fig. 9-6 contoured into the dashed curve.)

Serial versus Parallel

We can send and receive binary digital data one bit at a time along a single line or channel using a mode called *serial data transmission*. We can send and receive data at much greater speeds, however, by using multiple lines or a wideband channel, sending independent sequences of bits along each line or subchannel, a mode known as *parallel data transmission*.

A *parallel-to-serial* (P/S) *converter* receives bits from multiple lines or channels, and then retransmits them one at a time along a single line or channel. A *buffer* stores the bits from the parallel lines or channels, placing the bits in a *queue* for controlled, slower transmission along the serial line or channel. A *serial-to-parallel* (S/P) *converter* takes incoming bits from a serial line or channel, and re-sends the bits out in batches along multiple parallel lines or channels. Figure 9-7 illustrates a communications link using a P/S converter at the source and an S/P converter at the destination.

Digital Signal Processing

An electronic technology called *digital signal processing* (DSP) can improve the precision of digital data. The process enhances the quality of marginal or weak signals, so that the receiving equipment makes fewer errors. Binary digital signals have well-defined patterns that computers can easily recognize and clarify.

A DSP system minimizes or eliminates confusion between digital states that can result from electrical noise. In Fig. 9-8A, we see a hypothetical "noisy" binary digital signal before DSP. At B, we see the same signal after DSP has "cleaned it up."

TIP ***The DSP circuit makes logic-state "decisions" at defined, constant time intervals equal to a the duration of a bit. If the average strength of the incoming signal exceeds a certain level over the time interval, the output equals 1. If the average signal strength remains below another defined critical level for a certain time interval, the output equals 0.***

Digital Color

Our eyes and brains perceive *color* as a function of three independent radiant-light characteristics. When much of the energy in a ray or beam of visible light exists near a single wavelength, we see an intense *hue*. We define the vividness or "purity" of a hue as the *saturation*. We define the *brightness* of a particular hue as a function of how much total energy the light contains at the wavelength representing that hue. We can obtain a light ray or beam having any imaginable hue,

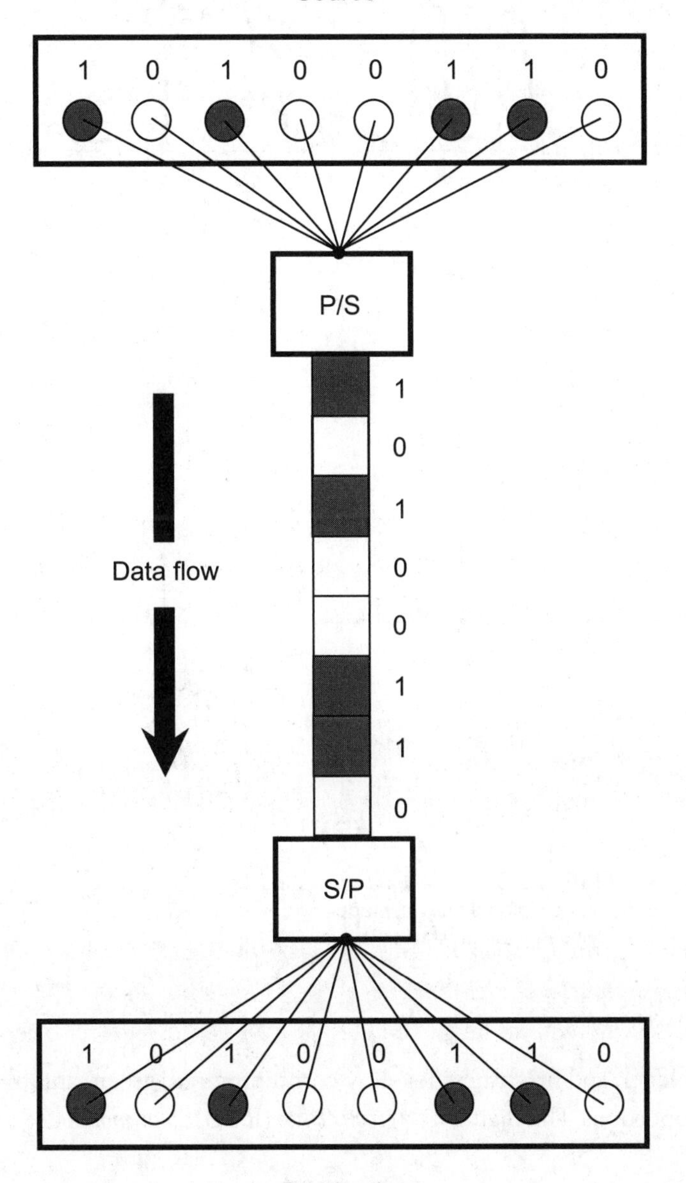

FIGURE 9-7 • A communications circuit employing parallel-to-serial (P/S) conversion at the source, and serial-to-parallel (S/P) conversion at the destination.

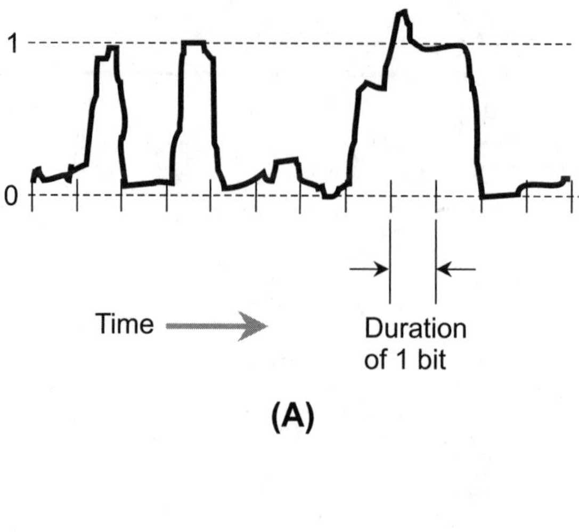

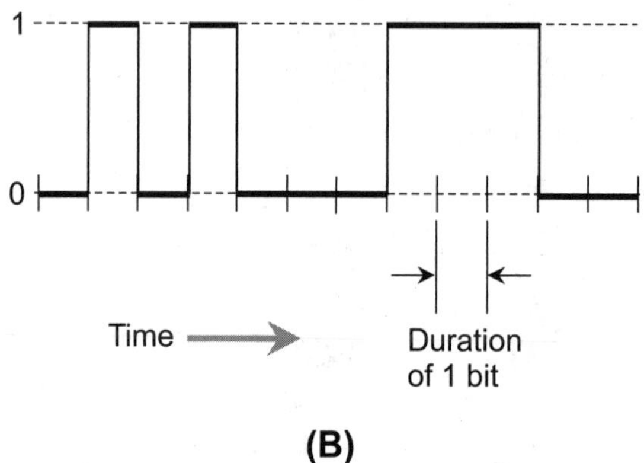

FIGURE 9-8 • Digital signal processing (DSP) can "clean up" a noisy digital signal (A), producing an output with well-defined logic states (B).

saturation level, and brightness level by combining red, green, and blue light in various proportions. The digital *red/green/blue* (RGB) *color model* takes advantage of this fact.

We can generate a color *palette* by combining pure red, green, and blue in various ratios, assigning each primary color an axis in rectangular three-dimensional (3D) space as shown in Fig. 9-9. We call the axes R (for red), G (for green), and B (for blue). The color intensity along each axis can range from 0 (the complete absence of energy) to 255 (the greatest possible amount of energy that the system can provide). The equivalent range in binary numeration goes from 00000000 to 11111111. When we combine

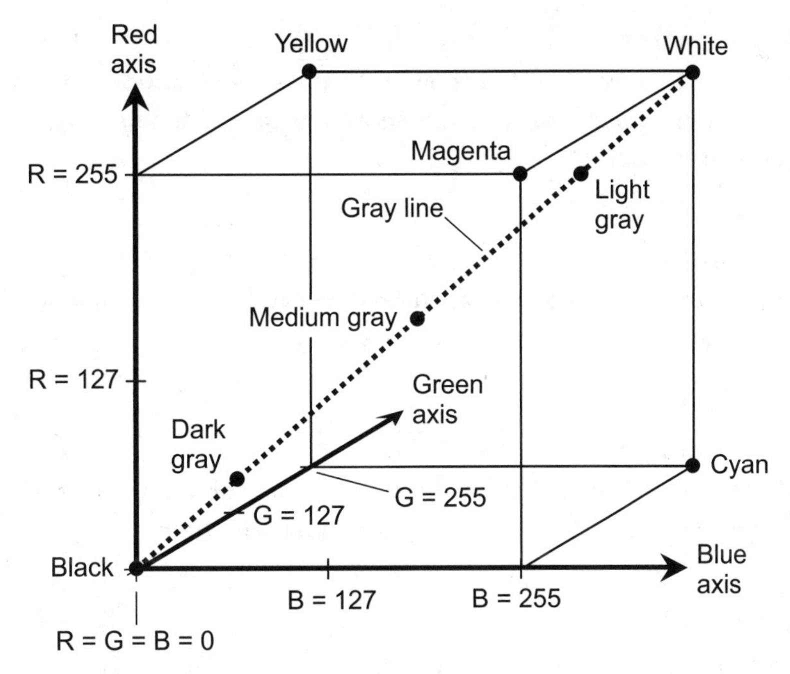

FIGURE 9-9 • In the RGB color model, visible-light colors can be represented as points inside a cube.

all three color axes into a "3D color space" with 256 possible brightness values along each axis, we obtain 256 × 256 × 256, or 16,777,216, possible colors. We can represent each color as a unique point inside the geometric cube defined by the three perpendicular axes.

Still Struggling

Some engineers represent individual colors in the RGB model as six-digit hexadecimal numbers such as 005CFF. In this scheme, the first two digits represent the red (R) intensity in 256 levels ranging from 00 to FF. The middle two digits represent the green (G) intensity, again in 256 levels ranging from 00 to FF. The last two digits represent the blue (B) intensity, once again in 256 levels ranging from 00 to FF.

PROBLEM 9-6

Suppose that we want to list all of the possible conditions that an n-bit binary digital signal can attain, where n is a positive integer. What general rule can we use?

SOLUTION

We can count upward in binary form from 0, writing every binary numeral as a sequence of n digits, until all of the digits equal 1.

PROBLEM 9-7

Using the rule described in the solution to Problem 9-6, list all of the conditions that a 2-bit binary digital signal can attain.

SOLUTION

See Table 9-7.

TABLE 9-7 Solution to Problem 9-7.

First bit	Second bit
0	0
0	1
1	0
1	1

PROBLEM 9-8

Using the rule described in the solution to Problem 9-6, list all of the conditions that a 3-bit binary digital signal can attain.

SOLUTION

See Table 9-8.

TABLE 9-8 Solution to Problem 9-8.

First bit	Second bit	Third bit
0	0	0
0	0	1
0	1	0
0	1	1
1	0	0
1	0	1
1	1	0
1	1	1

PROBLEM 9-9

Using the rule described in the solution to Problem 9-6, list all of the conditions that a 4-bit binary digital signal can attain.

SOLUTION

See Table 9-9.

TABLE 9-9 Solution to Problem 9-9.

First bit	Second bit	Third bit	Fourth bit
0	0	0	0
0	0	0	1
0	0	1	0
0	0	1	1
0	1	0	0
0	1	0	1
0	1	1	0
0	1	1	1
1	0	0	0
1	0	0	1
1	0	1	0
1	0	1	1
1	1	0	0
1	1	0	1
1	1	1	0
1	1	1	1

QUIZ

You may refer to the text in this chapter while taking this quiz. A good score is at least 8 correct. Answers are in the back of the book.

1. **Which of the following binary numbers is equivalent to the decimal 129?**
   A. 1111111
   B. 11111111
   C. 1000001
   D. 10000001

2. **Which of the following decimal numbers is equivalent to the hexadecimal F23?**
   A. 3875
   B. 1523
   C. 251
   D. 20

3. **In the hexadecimal number system, what number follows 999?**
   A. 1000
   B. 99A
   C. A000
   D. A99

4. **Suppose that we design an electronic music system using 9-digit binary numbers to represent various audio frequencies. Each binary number represents one, but only one, particular frequency. How many possible frequencies can such a system represent?**
   A. 64
   B. 512
   C. 1024
   D. 2048

5. **Figure 9-10 illustrates a circuit built of logic gates. Tables 9-10A through 9-10D show four possible representations for the states in this circuit. Which of these four tables correctly represents the circuit states at all five points *X, Y, P, Q*, and *R*?**
   A. Table 9-10A
   B. Table 9-10B
   C. Table 9-10C
   D. Table 9-10D

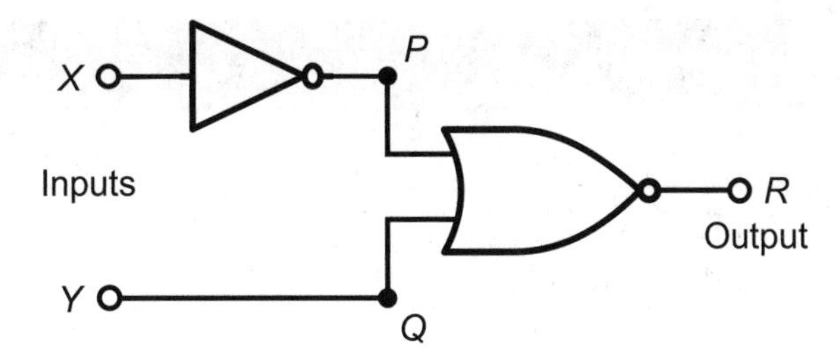

FIGURE 9-10 • Illustration for Quiz Question 5.

TABLE 9-10 Possible solutions to Quiz Question 5.

		A		
X	*Y*	*P*	*Q*	*R*
0	0	0	0	0
0	1	0	1	0
1	0	1	0	0
1	1	1	1	1
		B		
X	*Y*	*P*	*Q*	*R*
0	0	1	0	1
0	1	1	1	0
1	0	0	0	1
1	1	0	1	1
		C		
X	*Y*	*P*	*Q*	*R*
0	0	1	0	0
0	1	1	1	0
1	0	0	0	1
1	1	0	1	0
		D		
X	*Y*	*P*	*Q*	*R*
0	0	1	0	0
0	1	1	1	0
1	0	0	0	0
1	1	0	1	0

TABLE 9-11 Logic states for Quiz Question 6.

W	*X*	*Y*	*Z*	Output
0	0	0	0	1
0	0	0	1	1
0	0	1	0	1
0	0	1	1	1
0	1	0	0	1
0	1	0	1	1
0	1	1	0	1
0	1	1	1	1
1	0	0	0	1
1	0	0	1	1
1	0	1	0	1
1	0	1	1	1
1	1	0	0	1
1	1	0	1	0
1	1	1	0	0
1	1	1	1	0

6. Table 9-11 outlines all possible logic states for a digital circuit with four inputs *W, X, Y,* and *Z*. Figure 9-11 shows four different arrangements of logic gates. Which circuit in the figure produces the logic states in the table?

A. Figure 9-11A
B. Figure 9-11B
C. Figure 9-11C
D. Figure 9-11D

7. Figure 9-12 illustrates two electronic circuits. If we think of them as black boxes, how do their logic functions compare?

A. For any given set of inputs, the output from black box A is opposite the output from black box B.
B. The output from black box A is high if and only if the inputs differ, but the output from black box B is high if and only if the inputs are the same.
C. The output from black box A is high if and only if the inputs differ, but the output from black box B is high if and only if both inputs are high.
D. For any given set of inputs, the output from black box A is the same as the output from black box B. In other words, the two black boxes perform identical logic functions.

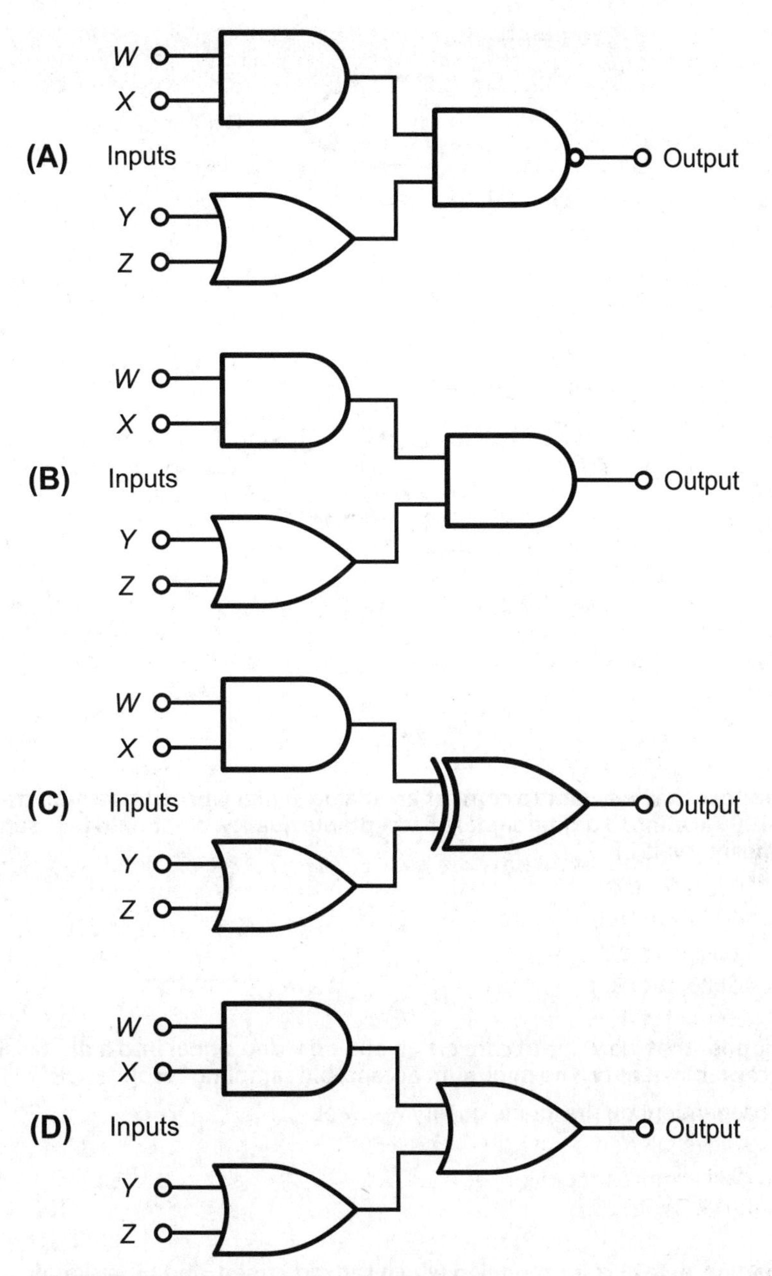

FIGURE 9-11 • Illustration for Quiz Question 6.

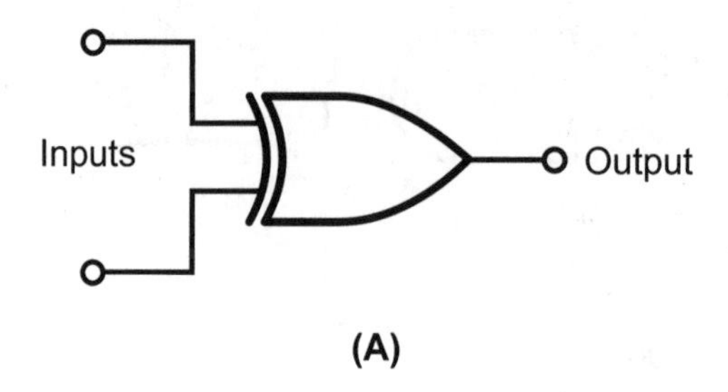

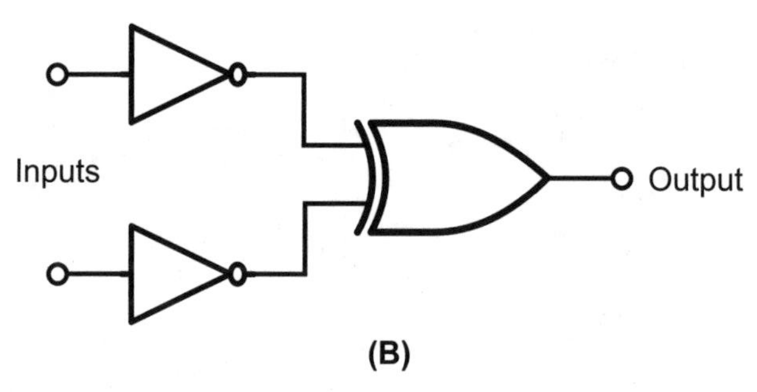

FIGURE 9-12 • Illustration for Quiz Question 7.

8. **Suppose that we want to convert an analog signal whose maximum frequency is 1000 kHz into a digital signal of acceptable quality. We should take samples at time intervals of**
   A. 1000 μs or less.
   B. 500.0 μs or less.
   C. 1.000 μs or less.
   D. 0.5000 μs or less.

9. **Suppose that we want to convert an analog video signal into a digital signal of acceptable quality. The minimum acceptable sampling resolution is**
   A. dependent on the image quality we want.
   B. 32 levels.
   C. 1024 levels.
   D. 1,048,576 levels.

10. **Imagine an RGB color model in which the red, green, and blue signals can each attain 32 discrete levels. This model can uniquely represent**
   A. 1024 different colors.
   B. 4096 different colors.
   C. 32,768 different colors.
   D. 65,536 different colors.

chapter 10

Reality Remystified

When we let logic team up with our imaginations to scrutinize the "real world," notions arise that can seem illogical. In this chapter we'll explore some vagaries of time, matter, space, and chaos. Get ready for a "mind journey"! Don't worry about taking a quiz at the end of this chapter, because you won't find one.

CHAPTER OBJECTIVES

In this chapter, you will

- See how time frames can displace and stretch.
- Learn how to travel into the future.
- Try to find the smallest material particle.
- Explore "space" having more than three dimensions.
- Discover why chaos prevails so often.
- Scrutinize the illogic of randomness.

The Illogic of Time

According to popular lore, Albert Einstein once declared, "God may be sophisticated but he is not malicious." I think Einstein wanted to suggest that the Creator didn't make the universe difficult to understand simply to spite us human beings or to assault our faith in logic. Nevertheless, some people can't believe the conclusions that Einstein reached, despite the fact that he pursued his theories with mathematical rigor.

The Light-Beam Conundrum

Sound waves could not propagate from place to place if there existed no air, water, or other material medium to carry them. If the source of an acoustic disturbance remains stationary with respect to the surrounding air, the sound waves travel outward at equal speeds in all directions relative to the source. If the source moves through the surrounding air, the sound waves travel outward faster in some directions than in others. Acoustic waves always propagate at a constant speed with respect to the *medium* (air), but not always at the same speed with respect to the *source*.

In the late nineteenth century, scientists believed that light waves in space behave as sound waves do in the earth's atmosphere. This assumption had no foundation other than intuition, which, as you ought to know by now, can lead brilliant minds astray. By assuming that light waves propagate at a constant speed with respect to the "fabric of space," scientists concluded that the universe must possess an absolute standard of motion. An object "at rest in the universe" would emit or intercept light rays at equal speeds in all directions. An object "moving through the universe" would emit or intercept light rays at speeds dependent on the direction.

With the foregoing notions in mind, astronomers set out to determine the absolute standard for motion in the universe. The basis for such an experiment seemed simple enough: Capture the light from stars in various parts of the sky and measure the arrival speeds of the rays. Rays would presumably come in fastest from stars located in the direction of the earth's motion, and slowest from stars in precisely the opposite direction. When the scientists compiled their results, however, they got a nasty surprise: Light beams arrived at the same speed from every single star, regardless of its location in the sky!

TIP ***No scientist could force herself to believe that the earth remains absolutely stationary with respect to the universe at large. That theory had gone out with Galileo Galilei, Johannes Kepler, and Isaac Newton. The only logical conclusion: Light beams in space travel differently than sound waves in the air. Albert Einstein, taking note of the experimental results, regarded the constancy of light speed as a sort of "astrophysical axiom."***

Synchronize Your Watches!

The notion of *simultaneity*—the condition wherein two or more events occur at the same time—seems simple at first thought. We might say that two events occur simultaneously when, as far as we can tell, they occur at the same instant. Of course, true simultaneity can have meaning only to the extent of our ability to measure time; two things that seem to take place at the same moment according to our body senses might not seem that way when we employ a precision chronometer such as an atomic clock. (The esoteric among us might argue that time itself can't exist, because instants in the mathematical sense constitute "time-points" with zero duration! But let's not get into that discussion now.)

We can always set two clocks so that they show exactly the same time, at least to the best of our ability to determine. Suppose, for example, that you set your wristwatch from a location near the time-standard shortwave-radio broadcast station WWV in Fort Collins, Colorado. Then you travel far away from that place, say to Sydney, Australia, and listen to the signals on a radio receiver. The signals from the broadcast station won't precisely agree with your watch any more—not because of any peculiarity involved with the motion of the traveling process, and not because your watch has lost its accuracy, but because of the finite speed at which the radio signals propagate around the globe. All electromagnetic waves travel through space or through the earth's atmosphere at approximately 186,000 miles per second (300,000 kilometers per second). For the signal from the broadcast station to get from Colorado to Australia, it takes approximately 0.07 seconds. If you set your watch by station WWV while you are in Colorado and then travel to Australia, you'll sense the difference.

Does WWV tell the correct time as you hear it, or does your watch show the correct time as you look at it? That's a "loaded" question, because you can use either time standard. Your perception of time depends on where you are relative to the time sources you consult. When two distinct events occur, we might reasonably call them "simultaneous" if and only if they *appear* simultaneous as we observe them. In the foregoing example, one event is the click of an atomic clock in Colorado, and the other event is the tick of your watch at some place far away. You can reset your watch to agree with WWV as you tour the town of Sydney, but when you return to Colorado, you'll find that a discrepancy of about 0.07 second has cropped up again.

We can show infinitely many examples of the above-described phenomenon. The moon orbits at a distance of roughly 1.3 light seconds from the earth. Visible light or radio waves therefore take approximately 1.3 seconds to traverse this distance through space. Nothing, as far as we know, can move any faster than electromagnetic fields through space. If we wish to say that an event occurs at the same time on the moon as it does on the earth, we must qualify our observation by specifying the point of reference. The situation becomes more complicated if we have many events that occur at widely scattered points in space. In such instances we can always define apparent simultaneity at one particular spot, but if we move somewhere else, we must revise our definition.

When Is a Second Not a Second?

Motion, as well as physical separation, complicates our perception of time. Motion causes not only a displacement in time, but a change in the apparent rate at which time progresses. An obvious manifestation of this phenomenon occurs with the so-called *Doppler effect*, named after the scientist who discovered it. Sound waves provide an everyday example.

As a train goes by and honks its horn, you hear the pitch of the horn go down. The wavefronts compress as the train engine comes toward you, and stretch out as the engine passes and moves away from you. The frequency, or number of acoustic vibrations per second, of the sound changes as the waves arrive at your ears. The same thing happens with electromagnetic waves, although not to a noticeable extent at everyday speeds. The Doppler shifting of light beams constitutes a form of "time distortion." At extreme speeds, the extent of this distortion can reach considerable proportions.

Doppler effect is not the only way in which extreme speeds "distort" time. Another effect, known technically as *time dilation*, was originally predicted by Einstein and verified by experimenters years later. At very high speeds, time appears to slow down from certain points of view, approaching a complete halt as the relative speed approaches the speed of light.

Imagine that we ride on a space ship equipped with a laser/sensor on one wall and a mirror on the opposite wall (Fig. 10-1). Imagine that the laser/sensor

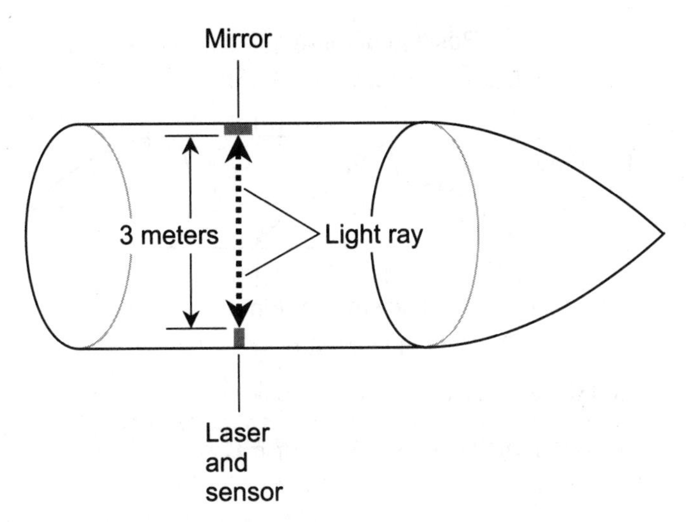

FIGURE 10-1 • A space ship equipped with a laser clock. This is what we see as we ride in the ship at constant speed, no matter how fast.

and the mirror are positioned so the light ray from the laser must travel perpendicular to the axis of the ship, and (once we get it moving) perpendicular to the ship's direction of motion. Suppose that we adjust the laser and mirror so they are separated by exactly 3 meters. Because the speed of light in air is 300 million meters per second, it takes 10 nanoseconds (10 thousand-millionths of a second) for the light ray to propagate across the ship from the laser to the mirror, and another 10 nanoseconds for the ray to return to the sensor. The ray therefore requires 20 nanoseconds to make one round trip from the laser/sensor to the mirror and back again.

Now imagine that we start up the ship's engines and get the vessel moving. We accelerate with the eventual goal of approaching the speed of light. Suppose that we accelerate to a sizable fraction of the speed of light, and then we shut off the engines so that the ship coasts through space. We measure the time it takes for the laser to go across the ship and back again. We move right along with the laser and the mirror, of course, so we find that the time lag is the same as it was when the ship was not moving; it's still 20 nanoseconds. This observation comes as no surprise to us, for it follows directly from Einstein's axiom. The speed of light has not changed—because it *can't*. The distance between the laser and the mirror has not changed either. Therefore, the laser beam's round trip takes the same length of time as it did before we got the ship moving. If we accelerate so the ship travels at 60%, then 70%, and ultimately 99% of the

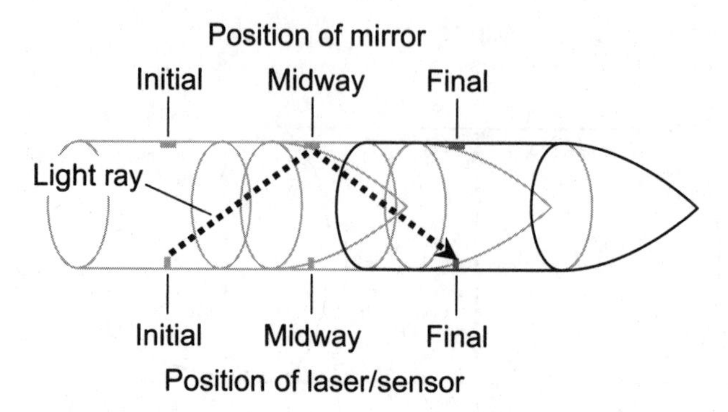

FIGURE 10-2 • This is what an earthbound observer sees as our laser-clock-equipped space ship moves at a sizable fraction of the speed of light relative to her reference frame.

speed of light, the time lag will always equal 20 nanoseconds as measured from a *reference frame*, or point of view, inside the ship.

Now imagine an observer sitting comfortably in an observatory on earth. She has a massive, powerful telescope that allows her to see inside our ship as it whizzes along through space. Imagine that she can see the laser, the mirror, and even the laser beam itself. Figure 10-2 shows what she sees. The laser beam travels in straight lines at 300 million meters per second relative to her reference frame. (Remember, the speed of light always appears the same, regardless of the motion of the observer.) But as this observer sees it, the laser beam must travel more than 3 meters to get across the ship. The ship travels so fast that, by the time the ray has reached the mirror from the laser, the ship has moved a significant distance forward. The same thing happens as the ray returns to the sensor from the mirror. From this observer's point of view, the laser beam will need more than 20 nanoseconds to travel across the ship and back.

Our laser-mirror-sensor device constitutes a form of "optical clock," based on the absolute constancy of the speed of light. We might say that it constitutes a perfect clock, based on an absolute universal constant (the speed of light). Yet, this clock gives a different reading from outside the ship as compared with inside. As the ship goes by, the rate of time progression inside it appears to slow down as seen from a "stationary" point of view. But inside the ship, time appears to go by at normal speed. As the ship's speed increases, so does the time-rate discrepancy. As the speed of the ship approaches the speed of light, the so-called *time dilation factor* grows without limit.

Traveling into the Future

In theory, time dilation allows astronauts to travel great distances without having to worry about not reaching remote destinations within their lifetimes. The nearest star, *Proxima Centauri*, is over four light years away from our solar system; it would take almost 9 years to complete a round trip to this star at the speed of light. But that's 9 years as earthbound people would measure it. If we could accelerate to a high enough speed, we could shorten the journey considerably for ourselves. In theory, no limit exists to how much we could compress the time. In practice, we would find ourselves constrained by the maximum thrust we could get from our space ship's engines, and also by the maximum amount of acceleration that our bodies could endure.

In his book *Cosmos*, Carl Sagan takes us into an imaginary space ship capable of accelerating indefinitely at one gravity, or about 10 meters per second squared. This kind of acceleration—exactly what we experience as we stand on the surface of our planet—would provide artificial gravity and, after a period of time, result in near-relativistic speed. It would be possible to travel anywhere in the known universe within the span of a lifetime.

Relativistic time dilation literally allows us to travel into the future! Suppose, for example, that we take a round trip to *Proxima Centauri*, which is 4.3 light years from us, and spend a few days there. If our average speed equals nine-tenths of the speed of light, we'll need 9 years and 7 months to complete the journey according to our earthbound colleagues. But according to our own reckoning, the trip will take about 4 years and 2 months. We'll propel ourselves into the future by the difference—5 years and 6 months.

Now suppose that we travel to a star 100 light years away in a ship that moves at an average of 99% of the speed of light. The round trip will take 101 years according to earth time, but we'll age only a little more than 14 years. Upon our return to the earth, we will emerge from our space vessel to see our planet nearly 87 years "older" than it "should be." We might return to find that our children had died, and perhaps that the world had become an unfamiliar place indeed.

TIP ***In the extreme, human beings might someday be able to travel so far into the future that they wouldn't have any idea what to expect upon their return to the home world. If people were to go to the farthest parts of our universe, they would return many millions of years from now. Even the sun might have died by then.***

Traveling into the Past

Some theorists imagine that we might travel backward time if we could build a spacecraft capable of traveling faster than the speed of light. Because time slows down and almost comes to a stop as we approach the speed of light, it's tempting to speculate that time would reverse its direction if we could exceed the speed of light. This extrapolation can serve as an example of a logical fallacy often committed by people who ought to know better! The fact that we approach a certain condition as we near a critical point does not logically imply that the condition will actually manifest itself if we reach the critical point, much less that the condition might be exceeded should we go beyond the critical point. When a mathematical function "blows up" to a *singularity* as the relativistic time-dilation factor does at the speed of light, the equations tell us absolutely nothing.

Relativity theory seems to suggest that time approaches a standstill as the speed of an object becomes arbitrarily close to the speed of light. However, we can't conclude from the mathematics of relativity that time will *stop* at the speed of light or *go backwards* at higher speeds. Have you ever tried to divide by zero? For a positive real-number variable x, the value of $1/x$ grows without limit as x approaches zero, but mathematicians do not define the value of 1/0. We might want to believe that 1/0 equals "infinity," but we can't verify the truth of such a proposition unless we can prove it. When we plug the speed of light into relativistic time-dilation equations, we end up in effect attempting to divide 1 by 0.

Skeptics of science-fiction time-travel scenarios have used *reductio ad absurdum* to prove that humans can't travel backward in time. The argument proceeds as follows. Suppose that we *can* do it. In that case we must commit some specific, voluntary act—execute some task at will—in order to make the journey. Let's call this act A. Whatever the technical details of A might be, we can prevent ourselves from doing A, because A is voluntary; we don't *have* to do it. Now imagine that we commit A, causing us to travel back in time by a few minutes. We find ourselves at a moment in time prior to our original commission of A, and then we decide not to commit A. Therefore, the time journey never takes place. But we just completed it! That's absurd, proving the falsity of our original proposition that we can travel backward in time.

The Twin Paradox

Let's return to the more realistic aspect of time travel: relativistic time dilation, which theoretically allows people to travel into the future. Consider two identical twins born on the same day. One of them takes a long space journey at relativistic speed. When the space traveler returns, the spacefaring twin is younger than the stay-at-home twin. The extent of the difference depends on the average speed of the traveler and the duration of the journey. The longer the trip at a given average speed, the greater the age difference at the end; the greater the average speed for a trip of a given duration, the greater the age difference at the end.

Now imagine twin brothers riding in space ships somewhere in the interstellar void where no reference point exists that they can call "stationary." The twins can define motion only with respect to something "fixed." Suppose that their ships hover close to each other, and that they can both look out their windows and agree that they're stationary relative to each other. The twins (let's call them Jim and Joe) synchronize their clocks so that they display the same date and the same time. What will happen if Joe fires up his rocket engines, speeds off to some distant point in the void, and then returns to Jim who waits without using ever having "moved"—that is, without ever having used his rocket engines—during that time? According to the special theory of relativity, Jim will see Joe's "time flow rate" slow down while Joe moves. When Joe returns, his clock will lag behind Jim's clock. Joe will be younger than Jim.

But why should we claim that Joe moves and Jim does not? They're both so far away from any meaningful point of reference that neither of them can come to any conclusion about absolute motion—can they? Suppose that Joe pulls away from Jim and goes a few light years (a light year equals about 4 million-million kilometers or 6 million-million miles) into space and then comes back, as shown in Fig. 10-3A. That's the way Jim sees the situation. But what happens if we take Joe's point of view instead? Can't we just as well say that Jim moved away from Joe in the opposite direction (Fig. 10-3B) and then came back? Relatively speaking, aren't these two models, one from Jim's point of view and the other from Joe's, identical? If so, then Joe sees Jim's "time flow rate" as moving slower than his own, so Joe should find that Jim's clock lags his own when they rendezvous at the end of their period of separation.

When confronted with this apparent paradox, some people declare that it disproves the special theory of relativity. An apparent contradiction exists: Joe is younger than Jim and Jim is younger than Joe, even as they sit together over a synthetic steak dinner and try to figure out what sort of madness motivated them to become space travelers in the first place. But we can't just trash Einstein's theory of relativity on the basis of this "thought experiment," because the validity of the theory has been demonstrated by real-life experimentation. Some scientists took an airplane aloft bearing an atomic clock. The experimenters

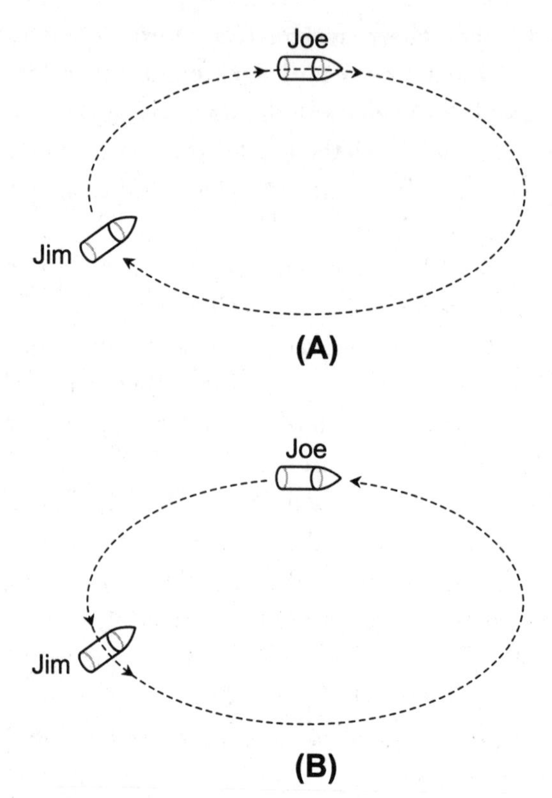

FIGURE 10-3 • In (A), Jim thinks Joe moves. In (B), Joe thinks Jim moves. Who is correct?

synchronized the airplane's clock with another atomic clock at a base on the earth's surface. When the plane returned, the experimenters found that the clock on the airplane lagged a tiny fraction of a second behind the clock at the base. The discrepancy exactly matched the value predicted according to the special theory of relativity.

PROBLEM 10-1

How might we try resolve the twin paradox?

SOLUTION

We can note that Joe uses his rocket engines, but Jim does not, and that fact constitutes a fundamental difference between the behaviors of the two space ships. Joe's velocity changes (he *accelerates*) while Jim's velocity remains constant (he does not accelerate). If you've taken a basic physics

course, you know that Joe feels the effect of his acceleration as a gravity-like force on his body inside the ship, while Jim senses no such force, remaining weightless all the time. Joe's motion therefore differs in an essential way from Jim's motion. This difference must contain the resolution to the paradox. In the foregoing discussion, we asked ourselves two questions:

- Can't we just as well say that Jim moved away from Joe in the opposite direction (Fig. 10-3B) and then came back?
- Relatively speaking, aren't these two models, one from Jim's point of view and the other from Joe's, identical?

The answers are "No!" and "No!"

PROBLEM 10-2

Once we've found a plausible way to debunk the twin paradox as described in the solution to Problem 10-1, how can we bring back the paradox by modifying the scenario, creating an apparently unresolvable problem once more?

SOLUTION

Imagine that Jim and Joe take identical "mirror-image" journeys, starting at and returning to a central, nonaccelerating point as shown in Fig. 10-4. Suppose that, at every instant in time (as a far-off observer would see the

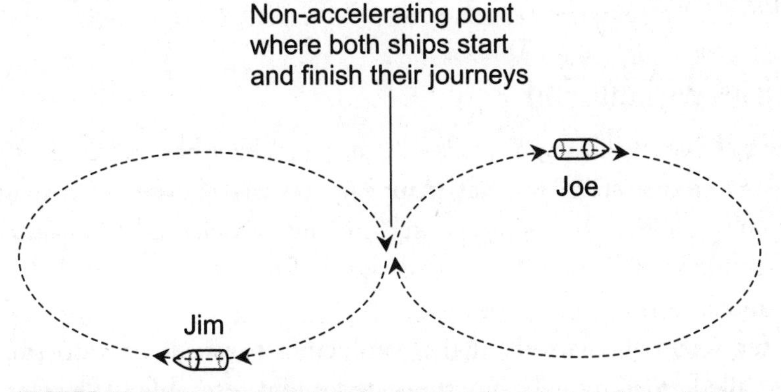

FIGURE 10-4 • We can reintroduce the twin paradox to account for acceleration. Here, Jim and Joe take "mirror-image" journeys.

situation), Jim's speed and acceleration are the same as Joe's, but in the opposite direction. We can no longer claim that Joe's motion differs qualitatively from Jim's. Their journeys are *congruent*. However, Jim still perceives Joe as moving, so Jim should age more than Joe during their travels; Joe still perceives Jim as moving, so Joe should age more than Jim. We've come back to the point of paradox.

Still Struggling

I can't explain away the paradox in the "solution" to Problem 10-2. Can you?

The Illogic of Matter and Space

The pure mathematician deals in theory. She needs only a pencil, some paper, a clear head, and a quiet place to think. The pure experimentalist, in contrast, makes observations and measurements, and leaves it up to others to draw conclusions from the resulting data. A professor once told me, "One experimentalist can keep a dozen theorists busy." This axiom certainly holds in physics if not in pure mathematics. The "real world" behaves in such an unreal way, when dissected by modern science, as to render it beyond most people's comprehension.

Particles without End

In the twentieth century, physicists began to realize that matter does not comprise continuous "stuff," the way it appears to a casual observer. Instead, matter is "grainy," made up of countless tiny, moving particles called *molecules*. Every conceivable kind of material thing consists of molecules so small that an ordinary microscope can't resolve them.

Later, scientists discovered that molecules break down into smaller particles called *atoms*, which they theorized might resemble little solar systems. A positively charged central *nucleus* shepherds a group of negatively charged particles called *electrons* into a set of orbital paths called *shells*. Some nuclei have a lot of positive charge, and many electrons orbit them. Some nuclei have less charge, and therefore hold correspondingly fewer electrons in orbit. Usually, the negative and positive electrical charges in an atom precisely balance, so that the whole atom carries a net electrical charge of zero. But some atoms have an excess of negative charge (an electron surplus) or an excess of positive charge (an electron shortage). Atoms can join together by sharing electrons.

The nuclei and electrons in an atom consist of smaller constituents known as *quarks*. In recent decades, particle physicists have begun to realize that matter is so complex that we might never finish the job of trying to figure out exactly how it's made. One of the most bizarre particles that theorists have discovered in their blackboards full of formulas is the *tachyon*. No one has actually isolated a tachyon yet, but these particles can supposedly travel faster than the speed of light, an impossibility in this universe according to the principles of relativity theory. *Neutrinos* also exist; they can penetrate the earth as easily as a beam of light can pass through a window pane. No doubt more particles—perhaps infinitely many more—await discovery.

A Mad Professor's Monologue

"Let's imagine," says Professor N., our mentor, "that the universe has infinitely many different kinds of particles."

"In that case, humans will never find them all," you say.

"Correct. Well, then, suppose that we scientists have deceived ourselves all these many decades, and only one type of material particle really exists, having zero volume and infinite density," says Professor N.

"What do you mean?" you ask.

"We know that the universe has a certain density," continues the professor. "Some dispute prevails among cosmologists as to the exact value of that density. But imagine, for the sake of discussion, that the density of our universe is equivalent to one mid-sized American-made car every thousand cubic miles."

"Okay," you say. "That's mighty weird, but I guess I understand it."

"We can break this down further," says N. "Suppose that each one of these imaginary cars weighs the same as 5000 baseballs. Then the universe has the density of 5000 baseballs per thousand cubic miles, or five baseballs per cubic mile," says Professor N.

"Okay," you say.

"If we keep breaking this down, say into marbles, with 20 marbles to the baseball, then we can say that the universe has a density of five times 20, or 100 marbles per cubic mile. It's still the same density as the universe made up of baseballs or cars, but we have a larger number of particles, each one of a smaller size."

"Keep going," you say.

"Suppose," says Dr. N., "that there are three lead shot to a marble, or 10,000 grains of sand to a marble. We might say that the universe has a density of 300 lead shot, or 1,000,000 grains of sand, per cubic mile, and we would still find ourselves talking about the same density for the universe overall."

"What are you getting at?" you ask.

"We can continue this process indefinitely," says Professor N. "We can break things down into bits of dust, molecules, atoms, and so on, all the way down to elementary particles—if such a thing exists—and have the overall cosmic density remain the same. But ..."

"The universe does not have uniform density," you say.

"Correct," says the professor. "Matter tends to 'clump' or 'congeal' into galactic clusters, individual galaxies, stars, planets, asteroids, comets, and meteoroids. These objects consist of molecules and atoms. Protons, neutrons, and electrons make up the atoms. All of the protons, neutrons, and electrons in turn consist of quarks; and we mustn't forget the role that energy plays in the mass of the universe. We have photons, neutrinos, perhaps tachyons ..."

"So what?" you ask, by now growing rather impatient with the good professor.

"Imagine, if you will, that matter comprises *infinitely* many particles, all of them having *zero* volume and *zero* mass. Infinity times zero can come out as anything we want, you know. We thus live in a universe made up of infinitely many pieces of nothing."

Order from Randomness

Whether or not the foregoing idea represents the true structure of the universe, we know that a lot of particles exist, and they constantly move. Scientists first noticed the effects of particle movement when they looked through microscopes and saw small objects such as bacteria "jittering" for no apparent reason. They concluded that the "jittering" must result from multiple collisions with, or bombardment by, smaller, unseen particles. Eventually, scientists discovered that those particles were building-blocks of matter: molecules.

Molecules move faster as the temperature rises because the particles have more kinetic energy at higher temperatures as compared with lower temperatures. With countless particles moving around and bumping into each other all the time, we can never predict the point in space that a given particle will occupy at a given instant. The complexity of the situation defies any practical resolution. In theory, however, we might suppose that if we had a powerful enough computer, we could predict the position of any particle at any time, because the number of particles in the universe is finite—huge, but finite!

However, if *infinitely* many particles exist within any particular parcel of space, then we can never, ever hope predict where any one of them would be at a given time, no matter how powerful a computer we build. That task would require calculations of infinite complexity, or else it would take an infinite amount of time. We might have to resolve infinite series of infinite complexity. Even the tiniest error might precipitate a mistake of major proportions.

Whether or not particles move at random, we know that as the scale grows—as we look at bigger and bigger parts of the cosmos—we can find order more easily. Although we cannot, with current technology, say exactly where a certain molecule will be at a certain time, we can usually say where a tree, or a house, or a mountain will be. We can remain reasonably confident that all the atoms in a tree will not move an equal distance in a specific direction all at once.

You can say with almost perfect confidence that when you leave your desk to get a glass of water, the desk will be in the same place when you come back to it. On a larger scale, you can predict the earth's climatic seasons with certainty because the earth orbits the sun in a predictable way. You can say that the sun will have essentially the same size and temperature tomorrow as it is today (although astronomers might remind you that this simple rule does not hold true for all stars). Our Milky Way galaxy is even more stable than our sun. Although individual stars explode fairly often because so many of them exist in our galaxy, we need not pass our days in fear that the entire Milky Way will blow up any time soon.

The Ultimate State of Order

Size creates order, or at least the illusion of order within limited time frames. But what about the vastness of eternity? Two contrasting theories have received attention in the last century concerning the evolution of the universe: the *steady-state theory* and the *big-bang theory*.

According to the steady-state model, the density of the universe remains constant as time passes; things always have been and always will be about the same as they are now. If things appear to get more and more stable with increasing size (randomness having less and less observable effect) then this theory agrees quite well with what we humans see in our limited life spans. According to the proponents of the steady-state theory, ultimate "order" in the universe should exist on the largest scale.

The big-bang theory holds that the universe began approximately 14 thousand-million years ago in a violent explosion. Everything existed, according to this theory, within one point of space, or in a tiny particle of phenomenal density. For some reason this particle, called the *primordial fireball*, exploded and formed all the matter in the universe. The big-bang theory implies that the universe changes in size with time, constantly expanding as a result of the primeval explosion. If the theory represents reality, then the density of the universe must decrease as time passes. The beginning, incredibly hot and dense, must gradually give way to a cold, sparse, dark end—or, if enough gravitation exists among all the galaxies, to eventual contraction and collapse.

TIP ***Both of these theories define an ultimate state of order. The steady-state theory, envisioned by the astronomer Fred Hoyle and others, offers esthetic appeal because it implies an eternal and inalterable state of affairs. The big-bang theory, advanced by George Gamow and others including most astronomers today, has equal if not greater appeal because of its glory, its drama, and its apparent agreement with the opening words of religious books! According to either theory, the near-randomness of things and events on a small scale yields to an orderly, if perhaps violent, state of affairs on the ultimate scale.***

Hyperspace

Have you read or heard that space is *three-dimensional*? A mathematician will tell you that, in order to uniquely define the position of any point in space, we need to specify three independent numbers called *coordinates*. Two coordinates don't give us enough information; four coordinates give us more than enough (and maybe too much, because ambiguity can result).

Lay people rarely imagine that more than three physical dimensions exist, but a mathematician can easily imagine *hyperspace* having four, five, six, seven dimensions—or hundreds, or thousands, or infinitely many. Four-space would need four coordinates to uniquely determine the position of a point; five-space would require five coordinates, and so on.

Hypospace

The simplest two-space comprises a flat geometric plane. The surface of any smooth object, such as a sphere or cone, also constitutes two-space. Imagine a specific two-space universe on which little creatures carry out their lives. They're not like ants on a table or ball, but rather like "flat ants" (with zero thickness) who are mathematically confined to the surface of the table or ball. Obviously, such a constraint imposes severe limitations on the mode of existence! These creatures live not in hyperspace, nor even in ordinary space, but in *hypospace*. They're prisoners of geometry, at least from our point of view.

Imagine one of these two-dimensional creatures, imprisoned inside a rigid, unbreakable square (Fig. 10-5). She can't get out; she's imprisoned! Yet we can

FIGURE 10-5 • In two-space, a square could imprison a creature. We can see how to get her out, but she has no clue.

easily see how she might escape, if she only had access to the third dimension. We might reach into the square, grab the hapless prisoner, yank her off the two-dimensional surface that constitutes her universe, and plop her back down outside the square. We'd find the task easy, but she'd find it impossible to imagine, let alone do.

Now let's extend this idea to our own three-dimensional existence. Imagine that you're locked in a cube-shaped, unbreakable cell. You can't hope to escape. Nevertheless, a four-dimensional being, gazing upon you with compassion, could pull you out of the cell without touching, much less breaking, the walls, ceiling, or floor. If you're an ordinary three-space mortal like me, you can't directly envision how any hyperspace creature might do such a thing. Nevertheless, our inability to envision it in terms of three-dimensional logic doesn't prove it impossible!

Time-Space

Another, more comprehensible tactic exists by which you could escape from a cubical jail cell, assuming that you could travel at will both backward and forward through time. If you, endowed with such power, found yourself locked in a sealed, unbreakable chamber, you could move forward in time until the cell crumbled into dust. Then you could walk for a sufficient distance in any direction, go back in time to the "previous present," and find yourself outside the cell. Alternatively, you could zip back in time to some year before the structure was built, walk a good distance, go forward in time again to the "previous present," and revel in your freedom.

This theory seems elegant until you apply some simple logic. We've already shown that the notion of backward time travel gives rise to a direct contradiction. Here's another argument to that effect. Suppose that when you return to the "previous present" in either of the above-described scenarios, you make a slight miscalculation and arrive a minute before you left. You could look into the cell's window and see yourself inside getting ready to make the trip you just finished. You might shout to your "second self," who would respond with shock. You might then think back and recall that, prior to your temporal odyssey, *no one had greeted you from outside the cell!* You'd exist in two places at the same time, and yet, as you tried to unravel the mystery before you, realize that you could not exist right where you stood.

Time as a Dimension

Physicists and astronomers sometimes talk of time as the fourth dimension. But they also entertain the existence of a fourth geometric dimension. The evidence for this latter notion comes from observational evidence of a mathematical theory that three-space is "warped" or "curved." Just as a two-dimensional surface can be curved with respect to three-space (imagine the surface of the earth, for example), three-space can be curved with respect to four-space. Einstein predicted that astronomers would someday detect this curvature, and they did. In fact, we can theorize that three-space is "warped" (the technical term is *non-Euclidean*) not only at a few special points, but everywhere.

Astronomers base the notion of "warped" three-space on the theory's ability to explain observed facts. Einstein demonstrated that light coming from distant stars would "bend" in the gravitational field of the sun. The predicted extent of the curvature was not great, but sufficient that it ought to be observable with the aid of a telescope. When scientists finally did see the curvature of starlight passing near the sun during a total solar eclipse, the results agreed with Einstein's prediction.

Non-Euclidean three-space has some truly bizarre properties. For example, the shortest distance between two points in "warped" space is not necessarily a straight line. How would you get from your house to Moscow, Russia? The shortest path that you could take would constitute an arc, unless you were willing to do a lot of digging! In non-Euclidean three-space, the same principle applies.

The "curvature" of three-space suggests that a fourth geometric dimension exists. If any n-space (where n is any positive integer you want) is non-Euclidean, we can suppose that there exists an $(n + 1)$-space. A curved one-dimensional line needs two dimensions to exist, a curved two-dimensional surface needs three dimensions to exist, so a "curved" three-dimensional surface must need four dimensions to exist—right? We can extrapolate into more dimensions forever.

The question now comes to us: How many dimensions exist in the "ultimate cosmic reality"? Centuries ago, humans couldn't imagine more than three, but now we have evidence to the effect that four or more spatial dimensions exist. (Some cosmologists suggest that there could be many more than that.) Could four-space, too, constitute a non-Euclidean universe? We don't know, but we can propose a definition of n-space as follows:

- A geometric space has n dimensions if and only if it is possible to place exactly n straight rods with their ends touching so that each rod is perpendicular to all of the others.

We can't place four rods together this way in ordinary three-space. In a four-space we could, but few people can envision such an activity. We might think of time as a fourth dimension, and then we could consider time to "flow" in a direction "perpendicular" to three-space, and therefore perpendicular to all three rods placed together at right angles. We can use this principle along with computer animation software to make "time-space drawings" of objects that appear and disappear.

Still Struggling

Any rendition of five-space remains impossible for us to directly make, even with the addition of time as a dimension. As we increase the number of dimensions, the difficulty increases. Nonetheless, we can use the preceding definition to set up a mathematical coordinate system defining a hyperspace of any number of dimensions.

Infinity-Space

We can mathematically define a space having any natural number (nonnegative whole number) of dimensions, no matter now great. An interesting question follows: can we define *infinity-space*?

Suppose that we want to uniquely define a point in infinity-space. To do that, we must write out an ordered, infinite sequence of numbers, one for each axis in a coordinate system with infinitely many axes. We might try to denote such an "infinity-tuple" by writing

$$P = (x_1, x_2, x_3, x_4, x_5, \ldots)$$

where P represents a particular point. We could never name all of the values in this expression, no matter how much time we had or how fast we write them.

There exist infinitely many natural numbers, and we can never name them all; but mathematicians assure us that the set of natural numbers exists. Our inability to *name* every member of a set does not preclude the *existence* of the set. Therefore, if we imagine a point in infinity-space, we can say that a unique set of coordinates defining that point does exist, even if we can't name all the coordinates.

Even the simplest dimensional spaces involve an amazing degree of abstraction. We can never actually name all the points on a one-dimensional real-number line, but we still know that they all exist, don't we? The same holds for two-space, three-space, and so on. We take a leap of faith when we imagine a space of even a few dimensions.

TIP ***Why should we find it more difficult to imagine infinity-space than to imagine any other coordinate space?***

The Dwindling-Displacement Effect

Let's return to the idea of non-Euclidean space. Consider the surface of the earth, and the distance from your home town to Moscow, Russia—a few thousand kilometers if you live in the United States. Think of the surface of the earth as a spherical two-space, and think of the distance you must travel over the surface of that sphere to get from your driveway to the Kremlin. Then think of three-space, and the distance you would have to travel if you could dig a tunnel straight through the earth's interior. You wouldn't have to travel as far through the tunnel as you would over the surface.

The distance between any two points on a sphere, as measured in a "straight line" through the interior of the sphere, is always less than the distance between those same two points as defined over the surface. This same effect holds true for most points on any curved surface. Only a flat or Euclidean surface actually contains all of the shortest paths among points on itself. Let's extrapolate this notion into more than three dimensions. Imagine a "hyperspherical" n-space. The shortest path between any two distinct points in such an *n-sphere*, as measured on the $(n - 1)$-dimensional "surface," must be longer than the direct path through n-space from one point to the other.

If we assume that our universe is non-Euclidean, then the distance between any two points in four dimensions is less than the distance between those same two points in three dimensions, because "straight" lines in our universe aren't really straight at all! Now imagine that the four-space, in which our three-space resides, is also non-Euclidean. In that case, an even shorter path between two fixed points exists in five-space. We can go on with this progression forever if we suppose that the "ultimate cosmos" has infinitely many dimensions, all of them non-Euclidean.

PROBLEM 10-3

Consider again an "infinity-tuple" representing a point *P* in infinity-space, as follows:

$$P = (x_1, x_2, x_3, x_4, x_5, \ldots)$$

We can't write all of the values here, no matter how much time we have. Can we identify them all anyway?

SOLUTION

In most cases, we can't. However, in certain special cases we can define points in which the coordinate values occur in a known mathematical sequence. For example, we might specify the point

$$A = (1, 2, 3, 4, 5, ...)$$

where the values keep increasing by 1 in a simple *arithmetic sequence*. We might specify another point

$$G = (1, 1/2, 1/4, 1/8, 1/16, ...)$$

where each value equals half the previous one, so that the values follow a *geometric sequence*. Of course, the easiest points to define in infinity-space have coordinate values that are all identical, such as

$$H = (2, 2, 2, 2, 2, ...)$$

PROBLEM 10-4

Imagine two fixed points *P* and *Q* in non-Euclidean three-space, such that d_3 represents the length of the shortest possible path between them (called the *geodesic*). Let d_4 represents the geodesic between *P* and *Q* through four-space. Then d_4 is less than d_3. If the four-space is non-Euclidean, then a still shorter geodesic of length d_5 exists between two fixed points in five-space. If that five-space is non-Euclidean as well, an even shorter geodesic of length d_6 exists between *P* and *Q* in six-space. Now suppose that infinitely many dimensions exist, all of them non-Euclidean. In the most extreme scenario, what could happen to the length of the geodesic between *P* and *Q* as we evaluate it through more and more dimensions?

SOLUTION

The length of the geodesic between points *P* and *Q* would keep decreasing toward some specific *lower bound* as we measure it through more and more dimensions. In the most outlandish situation, that lower bound would equal zero. Then the two points *P* and *Q* would coincide.

The Illogic of Chaos

Have you ever noticed that events—especially dramatic or catastrophic events—tend to occur in bunches? A few decades ago, the engineer and mathematician *Benoit Mandelbrot* observed and analyzed this effect. His work gave birth to the science of *chaos theory*.

Was Andrew "Due"?

In the early summer of 1992, south Florida hadn't experienced a severe hurricane since Betsy in 1965. The area around Miami gets hit by a hurricane once every 7 or 8 years on the average, and an extreme storm once or twice a century. Was Miami "due" for a hurricane in the early nineties? Had the time come for a big blow? Some people said so. The year 1992 was no more or less special, in terms of hurricane probability, than any other year. In fact, as the hurricane season began in June of that year, experts predicted a season of below-normal activity, and that prediction proved accurate in general—with one big glitch.

On August 24, 1992, Hurricane Andrew tore across the southern suburbs of Miami and the Everglades like a cosmic weed whacker gone wild, and became the costliest hurricane ever to hit the United States up to that date.

Did Andrew's unusual intensity have anything to do with the lack of hurricanes during the previous two and a half decades? No. Did Andrew's passage make a similar event in 1993 or 1994 less likely than it would have been if Andrew had not hit south Florida in 1992? No. There could have been another storm like Andrew in 1993, and two more in 1994. Theoretically, there could have been a half dozen more like it later in 1992!

TIP ***Have you ever heard about a tornado hitting some town, followed 3 days later by another one in the same region, and 4 days later by another, and a week later by still another? Have you ever flipped a coin for a few minutes and had it come up "heads" a dozen times in a row, even though you'd normally have to flip it for days to expect such a thing to happen? Have you witnessed some vivid example of "event-bunching," and wondered if anyone will ever come up with a mathematical theorem that tells us why this sort of thing seems so common?***

Slumps and Spurts

Athletes such as swimmers and runners know that improvement characteristically comes in spurts, not smoothly with the passage of time. Figure 10-6

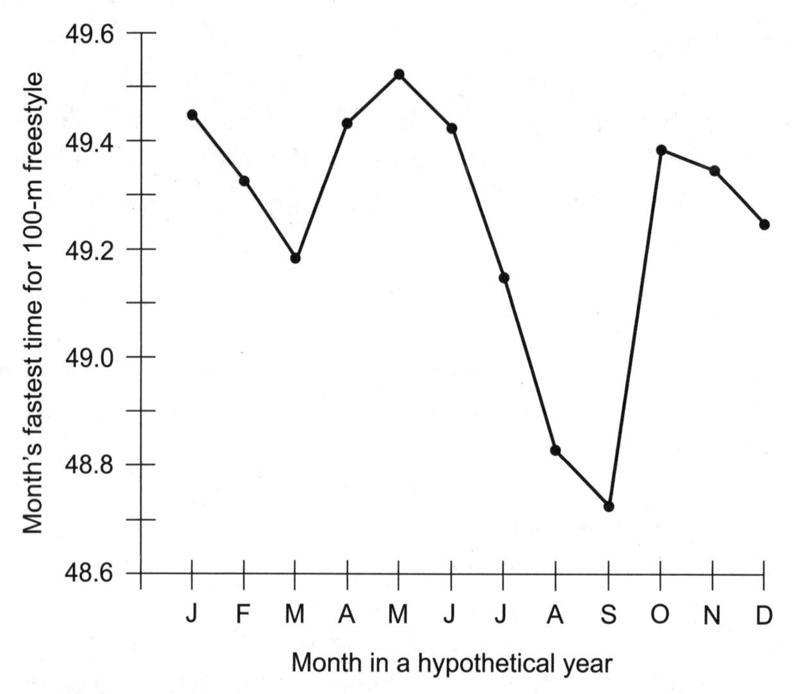

FIGURE 10-6 • A swimmer's monthly best times (in seconds) for the 100-meter freestyle, plotted by month for a hypothetical year.

illustrates an example in the form of a graph of the date (by month during a hypothetical year) versus race performance (as timed in seconds) for a hypothetical athlete's 100-meter freestyle swim. The horizontal scale shows the month, and the vertical scale shows the swimmer's fastest time in that month.

In this example, almost all of the swimmer's improvement occurs during June, July, and August. Another swimmer might exhibit performance that worsens during the same period of time. Does this irregularity mean that all the training done during times of flat performance represents time wasted? The swimmer's coach will say no! Why does improvement take place in sudden bursts, and not gradually with time? We observe similar phenomena in the growth of children, in the performance of corporate sales departments, and in the frequency with which people get sick.

Correlation, Coincidence, or Chaos?

Sometime during the middle of the twentieth century, a researcher noticed a strong correlation between the sales of television sets and the incidence of heart attacks in Great Britain. The two curves followed remarkably similar contours. The shapes of the graphs were, peak-for-peak and valley-for-valley, almost identical. Why?

Here's one theory: As people bought more television sets, they spent more time sitting and staring at the screens; the resulting idleness meant that they got less exercise; the people's physical condition therefore deteriorated, making them more susceptible to heart attacks. But this argument, if valid, couldn't explain the *uncanny exactness* with which the two curves followed each other, year after year. There would have been a lag effect if television-watching really did cause poor health, but no lag occurred. It was as if any hapless human could go out and buy a television set, and *instantly* expect to suffer ill effects.

Do television sets emit energy fields that cause immediate susceptibility to a heart attack? Is the programming so terrible that it causes immediate physical harm to viewers? Both of these notions seem "far-out." Were the curves obtained by the British researcher coincident for some unsuspected reason? Or did the whole thing constitute a complete coincidence lacking any explanation? Did no true correlation exist at all between television sales and heart attacks, a fact that would have emerged in time had the experiment continued for decades longer or involved more people?

TIP ***We may seek nonexistent cause-and-effect explanations for strange phenomena, getting more and more puzzled and frustrated as the statistical data keeps pouring in, demonstrating the existence of a correlation but giving no clue as to what might cause it. Applied to economic and social theory, this sort of "correlation-without-causation" phenomenon can lead to frightening hypotheses. Is another global war, economic disaster, or disease pandemic inevitable because that's "simply the way of the world"?***

Scale-Recurrent Patterns

Benoit Mandelbrot noticed that patterns recur over various time scales. Large-scale and long-range changes take place in patterns similar to those of small-scale and short-term changes. Events occur in bunches, and the bunches themselves take place in more protracted bunches following similar patterns. This effect exists in the increasing scale and in the decreasing scale.

Have you noticed that high, cirrostratus clouds in the sky resemble the clouds in a room where someone has recently lit up a cigarette? Or that these clouds look like the interstellar gas-and-dust clouds that make up diffuse nebulae in space? Patterns in nature often fit inside each other, as if the repetition of patterns over scale takes place on the basis of some principle "written in the operating manual for the universe."

You can observe a spectacular example of patterns that repeat by scale when you scrutinize the so-called *Mandelbrot set* using any of the various "zooming" programs available on the Internet. This geometric point set arises from a simple mathematical formula, yet it possesses infinite complexity. No matter how closely you magnify any part of the set, you see patterns that exhibit an eerie similarity at all scales.

The Maximum Unswimmable Time

If our hypothetical swimmer keeps training, how fast will he eventually swim the 100-meter freestyle? We already know that he can do it in a little more than 48 seconds. What about 47 seconds? Or 46 seconds? Or 45 seconds?

We can find obvious *lower bounds* to the time in which the 100-meter freestyle can be swum by a human. Any reasonable swim coach would agree that nobody will ever perform this event in 10 seconds. How about 11 seconds, then? Or 12 seconds? Or 13 seconds? Still ridiculous? How about 20 seconds? Or 25 seconds? Or 30 seconds? If we start at some impossible figure such as 10 seconds and keep increasing the number gradually, we will at some point reach a figure—let's suppose for the sake of argument that it's 41—representing the largest whole number of seconds too fast for anyone to swim the 100-meter freestyle.

Once we have two whole numbers, one representing a swimmable time (say 42 seconds) and the next smaller one representing an unswimmable time (say 41 seconds), we can refine the process down to the tenth of a second, and then to the hundredth, and so on indefinitely. There exists some value, exact to however small a fraction of a second we care to express it, representing the maximum unswimmable time (MUST) that a human being can attain for the 100-meter freestyle swim. Figure 10-7 shows an educated estimate (translation: wild guess) for this situation.

No one knows the exact MUST for the 100-meter freestyle, and we might argue that no human being (or even computer) can precisely determine it. But such a time nevertheless exists. How do we know that a specific MUST exists for the 100-meter freestyle, or for any other event in any other timed sport? A well-known theorem of mathematics, called the *theorem of the greatest lower bound*, makes it plain: "If there exists a lower bound for a set, then there exists a *greatest lower bound* for that set." A more technical term for "greatest lower bound" is *infimum*. In this case, the set in question is the set of "swimmable times" for the 100-meter freestyle. The lower bounds are the "unswimmable times."

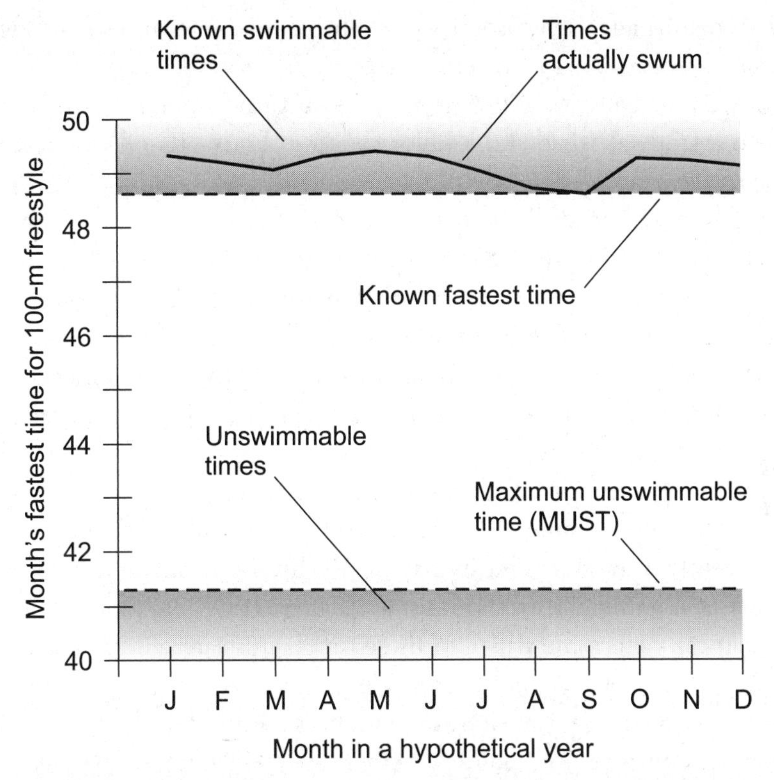

FIGURE 10-7 • According to the theorem of the greatest lower bound, there exists a maximum unswimmable time (MUST) for the 100-meter freestyle.

What's the probability that a human being will come to within a given number of seconds of the MUST for the 100-meter freestyle in, say, the next 10 years, or 20 years, or 50 years? Sports writers may speculate on it; physicians may come up with ideas; swimmers and coaches doubtless have notions too. But no one will ever know until after those 10, 20, or 50 years have passed.

The Butterfly Effect

The tendency for small events to have dramatic long-term and large-scale consequences has a catchy name: the *butterfly effect*. This expression arises from a hypothetical question: Can a butterfly taking off in China affect the development, intensity, and course of a hurricane six months later in Florida? At first, such a question seems outlandish. But suppose the butterfly creates a tiny air disturbance that produces a slightly larger one, and so on, and so on, and so on. According to butterfly-effect believers, the insect's momentary behavior could constitute the trigger that ultimately makes the difference between a tropical wave and a killer cyclone thousands of kilometers away and many days later.

We can never know all the consequences of any particular event. History happens once—and only once. We can't make repeated trips back and forth through time and let fate unravel itself over and over after "tweaking" this, or that, or the other little detail. But events can conspire, or have causative effects over time and space, in such a manner as to magnify the significance of tiny events in some circumstances. Scientists have programmed computers to demonstrate the phenomenon.

Suppose you go out cycling in the rain and subsequently catch a cold. The cold develops into pneumonia, and you barely survive. Might things have turned out differently if the temperature had been a little warmer, or if it had rained a little less, or if you had stayed out for a little less time? No practical algorithm exists that can determine which of these tiny factors are critical and which are not. But we can set up computer models, and we can run programs, that in effect "replay history" with various parameters adjusted. In some cases, certain variables exhibit threshold points where a tiny change right now will dramatically affect the distant future.

Scale Parallels

In models of chaos, patterns repeat in large and small sizes for an astonishing variety of phenomena. Compare, for example, the image of a spiral galaxy as viewed through a large telescope with the image of a hurricane as seen from an earth-orbiting satellite. The galaxy's stars compare to the hurricane's water droplets. The galaxy's spiral arms compare to the hurricane's rainbands. The eye of the hurricane is calm and has low pressure; everything rushes in toward it, as if it were a black hole. The water droplets, carried by winds, spiral inward more and more rapidly as they approach the edge of the eye. In a spiral galaxy, the stars move faster and faster as they fall inward toward the center, which may indeed contain a cosmic black hole.

Both pressure and gravitation can, as they operate over time and space on a large scale, produce the same general form of spiral. You'll find similar spirals in the Mandelbrot set and other mathematically derived patterns. The *Spiral of Archimedes* (a standard spiral easily definable in analytic geometry) occurs often in nature, and in widely differing scenarios. We can easily convince ourselves that these structural parallels represent something more than sheer coincidence—that a cause-and-effect relationship must exist. But what cause-and-effect factor can make a spiral galaxy in outer space appear and behave so much like a hurricane on the surface of the earth?

PROBLEM 10-5

Can the MUST scenario, in which a greatest lower bound exists, apply in a reverse sense? For example, might a *minimum unattainable temperature* (MUAT) exist for our planet?

SOLUTION

Yes. The highest recorded temperature on earth, as of this writing, is approximately 58°C (136°F). Given current climatic conditions, we can easily "invent" an unattainable temperature, for example, 500°C. We can then start working our way down from this figure. Clearly, 400°C is unattainable, as is 300°C, and also 200°C (assuming runaway global warming doesn't take place, in which our planet might end up with an atmosphere like that of Venus). What about 80°C? What about 75°C? A theorem of mathematics, called the *theorem of the least upper bound*, makes the situation plain: "If there exists an upper bound for a set, then there exists a *least upper bound* for that set." It follows that some MUAT for our planet must exist, given current climatic conditions.

The Malthusian Model

Some people grimly apply chaos theory to portray doomsday characteristics of the earth's population growth. Suppose we want to find a function that can describe world population versus time. The simplest model allows for an exponential increase in population, but this so-called *Malthusian model* (named after its alleged inventor, Thomas Malthus) fails to incorporate factors such as disease pandemics, world wars, or the collision of an asteroid with our planet.

The Malthusian model begins with the notion that the world's human population increases geometrically—in the same way that bacteria multiply—while the world's available supply of food and other resources can only increase arithmetically. It follows that a pure Malthusian population increase can only go on for a certain length of time. When we reach a certain critical point, the population will no longer increase, because the earth will become too crowded and there won't be enough resources to keep everyone alive for a normal lifespan. What will happen then? Will the population level off smoothly? Will it decline suddenly and then increase again? Will it decline gradually and then stay low? The outcome depends on the values we assign to certain parameters in the function that describes population versus time.

A Bumpy Ride

The limiting process for any population-versus-time function depends on the extent of the disparity between population growth and resource growth. If we consider the earth's resources finite, then the shape of the population-versus-time curve depends on how fast people reproduce until a catastrophe occurs. As the reproduction rate goes up—as we "drive the function harder"—the time period until the first crisis decreases, and the ensuing fluctuations become more wild. In the extreme case, the predictions become as dire as any science-fiction writer might conjure up in her most morbid funk.

Malthusian population growth takes place according to a specific formula. Statisticians, social scientists, biologists, mathematicians, and even some politicians have run this formula through computers for various values of population rate increase (called r or the *r factor*), in an attempt to predict what would happen to the world's population as a function of time on the basis of various degrees of "population growth pressure." It turns out that a leveling-off condition occurs when the value of r is less than about 2.5. The situation becomes more complicated and grotesque with higher values of r. As the value of the r factor increases, it "drives the function harder," and the population increases with greater rapidity until a certain point in time. Then chaos ensues.

According to computer models, when the r factor remains low, the world population increases, reaches a peak, and then falls back. Then the population increases again, reaches another peak, and undergoes another decline. This takes place over and over, but with gradually diminishing wildness. The catastrophes, however they might manifest themselves, become less and less severe. Mathematicians would say that a *damped oscillation* occurs in the population function as it settles down to a *steady state* (Fig. 10-8A).

In real life, humanity might keep the r factor low by means of extensive, worldwide public education programs. Conversely, the r factor could increase if all efforts at population control fail. Computers tell us with unblinking screens what they "think" will happen in that case. If r rises to a sufficient value,

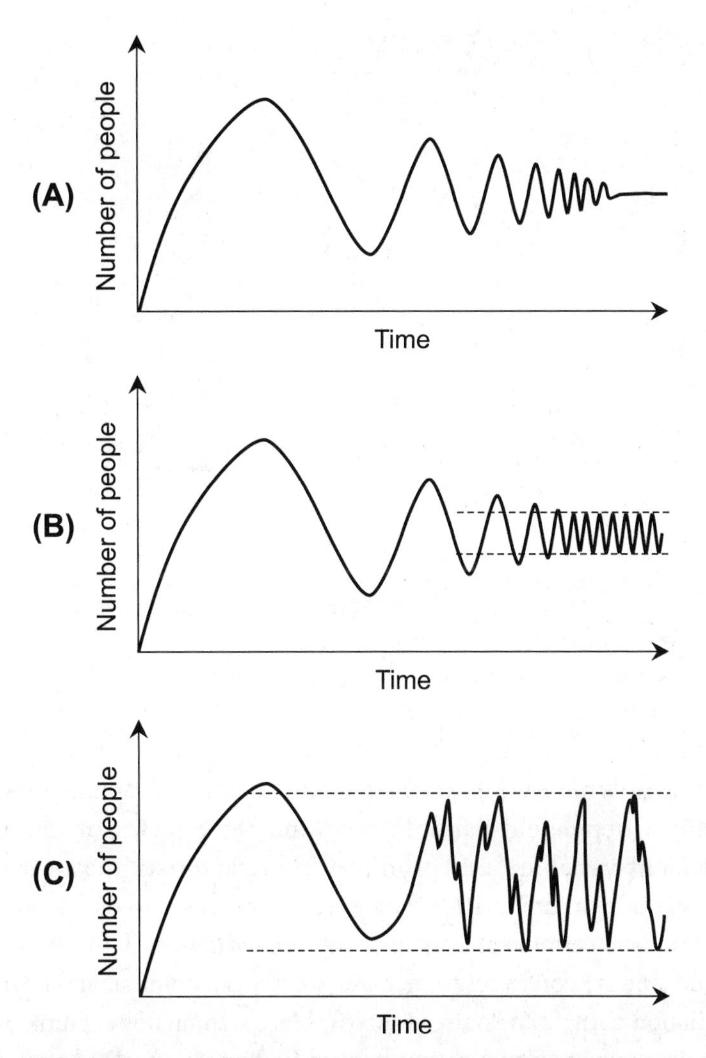

FIGURE 10-8 • Number of people in the world as a function of time. In (A), a small *r* factor produces eventual stabilization. In (B), a large *r* factor produces oscillation. In (C), a very large *r* factor produces chaotic fluctuation.

the ultimate world population will not settle down. Instead, it will oscillate indefinitely between limiting values as shown in Fig. 10-8B. The amplitude and frequency of the oscillation will depend on how large the *r* factor becomes. At a large enough critical value for the *r* factor, the population-versus-time function fluctuates crazily, never settling down to any apparent oscillation frequency, although the population "peaks and valleys" might remain within definable bounds (Fig. 10-8C).

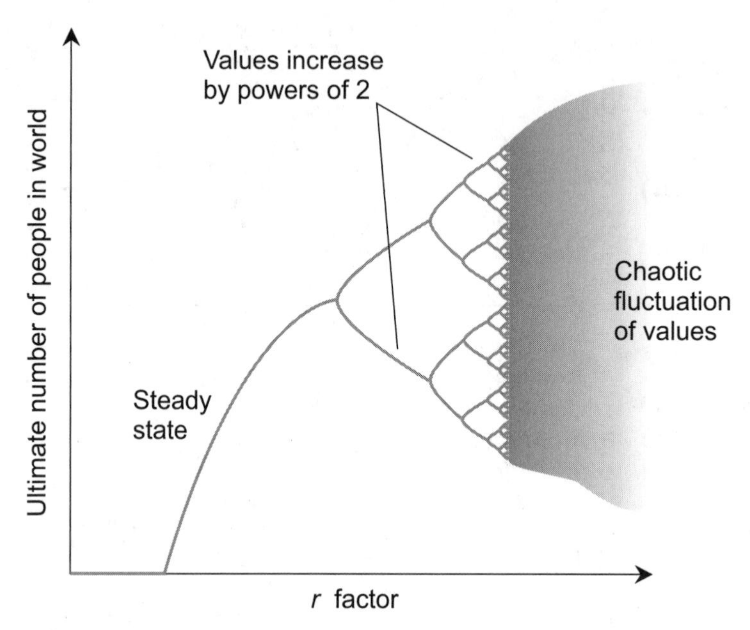

FIGURE 10-9 • Generalized, hypothetical graph showing "final" world population as a function of the relative *r* factor.

A graph in which we plot the world's ultimate human population on the vertical (dependent-variable) axis and the *r* factor on the horizontal (independent-variable) axis produces a characteristic pattern something like the one shown in Fig. 10-9. The function breaks into oscillation when the *r* factor reaches a certain value. At first this oscillation has defined frequency and amplitude. But as *r* continues to increase, we reach a point at, and beyond, which the oscillation turns into "noise" (chaos). Here's an analogy. Think about what happens when you ramp up the audio gain (volume control) of a public-address sound system until feedback from the speakers finds its way to the microphone, and the speakers begin to howl. If you increase the audio gain some more, the oscillations get louder. If you drive the system harder still, the oscillations increase in fury until, when you crank the gain up all the way to the top, the system roars like thunder.

TIP ***Does the final population figure in the right-hand part of Fig. 10-9 truly represent unpredictable variation between extremes? If the computer models represent reality, then the answer is yes. By all indications, the gray area in the right-hand part of Fig. 10-9 portrays a state of chaos. We can only hope that our world never enters a "gray area" like that.***

What Is Randomness?

Statisticians occasionally want to obtain sequences of values that occur at random. What constitutes randomness? Here's one definition that we can apply to single-digit numbers:

- A sequence of digits from the set {0, 1, 2, 3, 4, 5, 6, 7, 8, 9} is random if and only if, given any digit in the sequence, there exists no way to predict the next one.

At first thought, the task of generating a sequence of random numbers in this fashion seems easy. Suppose we chatter away out loud, carelessly uttering digits from 0 to 9, and record our voice on a computer disc? Won't that work? No! Everyone has a "leaning" or preference for certain digits or sequences of digits, such as 5 or 58 or 289 or 8827. If a sequence of digits proceeds in truly random fashion, then over long periods a given digit x will occur exactly 10% of the time, a given sequence xy will occur exactly 1% of the time, a given sequence xyz will occur exactly 0.1% of the time, and a given sequence $wxyz$ will occur exactly 0.01% of the time. These percentages should hold for all possible sequences of digits of the given sizes, and similar rules should hold for sequences of any length. But if you speak or write down digits for a few days and record the result, you can have almost complete confidence that things won't work out as such.

Here's another definition of randomness. We base this definition on the hypothesis that all artificial processes contain inherent orderliness:

- A sequence of digits is random if and only if there exists no algorithm capable of generating the next digit in a sequence, on the basis of the digits already generated in that sequence.

According to this definition, if we can show that any digit in a sequence constitutes a function of those before it, the sequence is not random. We can therefore rule out many sequences that seem random to the casual observer. For example, we can generate the value of the square root of 2 (or $2^{1/2}$) with an algorithm called *extraction of the square root*. We can apply this algorithm to any whole number that's not a perfect square. If we have the patience, and if we know the first n digits of a square root, we can find the $(n + 1)$st digit by means of this process. It works every time, we get the same result every time. The digits in the decimal expansion of the square root of 2, as well as the decimal expansion of any other irrational number, emerge in the same sequence—*exactly* the same sequence—every time a computer grinds it out. The decimal expansions of irrational numbers therefore do not give us random-digit sequences according to the above definition.

If the digits in any given irrational number fail to occur in a truly random sequence, where can we find digits that really occur at random? Does such a sequence even exist within the realm of "real-world thought"? If we can't generate a random sequence of digits using an algorithm, does this fact rule out any thought process that allows us to identify the digits? Are we looking for something so elusive that, when think we've finally discovered it, the very fact that we've gone through a thought process to find it proves that we have not? If that's true, then how can statisticians get hold of a random sequence that they can employ to their satisfaction?

In the interest of practicality, statisticians often settle for *pseudorandom* digits or numbers. The prefix *pseudo-* in this context means "pretend" or "for all practical purposes." Plenty of algorithms can generate strings of pseudorandom digits.

TIP ***You can search the Internet and find sites with information about pseudorandom and random numbers. The Web offers plenty of good reading, and numerous downloadable programs that can turn a home computer into a generator of pseudorandom digits. Go to your favorite search engine and enter the phrase "random number generator." But remember: Update your antivirus program before you download any executable file!***

Final Exam

Do not refer to the text when taking this exam. Strive to get at least 75 of your answers correct. The back of the book contains the answer key. You might want to have a friend check your score the first time you take this exam, so you won't "memorize" the answers to particular questions in case you want to try again.

1. **The most straightforward method of proving a theorem is known as**
   A. *reductio ad absurdum.*
   B. mathematical induction.
   C. intuitive derivation.
   D. inductive reasoning.
   E. deductive reasoning.

2. **Suppose we want to prove that a certain property holds true for all of the elements in a denumerably infinite set. Which of the following principles should we consider first?**
   A. The law of the contrapositive
   B. DeMorgan's law for conjunction
   C. The distributive law of disjunction over conjunction
   D. Mathematical induction
   E. *Reductio ad absurdum*

3. **We can express the law of excluded middle in propositional logic by writing one of the following symbol sequences. Which one?**
   A. $P \rightarrow (\neg P) = T$
   B. $P \wedge (\neg P) = T$
   C. $P \vee (\neg P) = T$
   D. $P \leftrightarrow (\neg P) = T$
   E. $P \rightarrow (\neg P) = F$

4. **Boolean multiplication represents the equivalent of propositional-logic**
   A. disjunction.
   B. negation.
   C. implication.
   D. conjunction.
   E. equivalence.

5. **Fill in the blank to make the following sentence true: "When we create a mathematical system, we accept __________ without definition or proof."**
   A. elementary terms
   B. definitions
   C. theorems
   D. corollaries
   E. implications

6. **Which, if any, of the following three statements A, B, or C can describe a system of binary logic?**
   A. Three truth values exist: "false," "neutral," and "true."
   B. Two truth values exist: "false" and "true."
   C. A continuum of truth values exists from "totally false" through "neutral" to "totally true."
   D. All of the above A, B, and C.
   E. None of the above A, B, or C.

TABLE EXAM-1 Table for Final Exam Question 7.

X	Y	X # Y
F	F	T
F	T	T
T	F	F
T	T	T

7. Table Exam-1 portrays the truth values for a certain operation in propositional logic, denoted with a pound sign (#). Which of the following symbols belongs in place of the pound sign?

A. An arrow pointing to the right (→).

B. A double-headed arrow (↔).

C. A wedge (∨).

D. An inverted wedge (∧).

E. None of the above.

8. We can define the boolean XNOR operation as the

A. conjunction of two variables, followed by negation of the result.

B. negation of a single variable, followed by conjunction.

C. negation of a single variable, followed by inclusive disjunction.

D. inclusive disjunction of two variables, followed by negation of the result.

E. exclusive disjunction of two variables, followed by negation of the result.

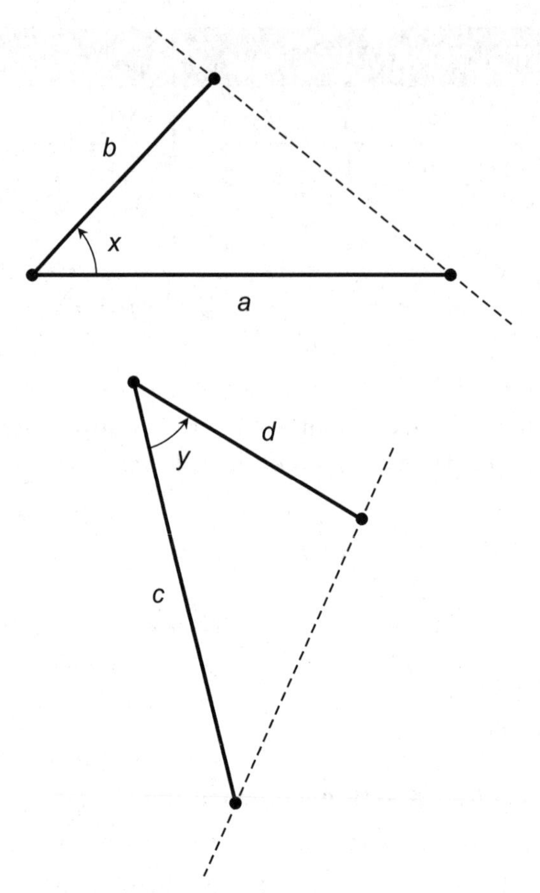

FIGURE EXAM-1 • Illustration for Final Exam Questions 9 and 10.

9. **In Fig. Exam-1, suppose $a = c$, $b = d$, and $x = y$. These facts make it possible to prove that the two triangles, formed by the vertex points shown, are**
   A. directly similar.
   B. inversely similar.
   C. directly congruent.
   D. inversely congruent.
   E. None of the above.

10. In Fig. Exam-1, suppose $a/c = b/d$ and $x = y$. These facts make it possible to prove that the two triangles, formed by the vertex points shown, are

A. directly similar.

B. inversely similar.

C. directly congruent.

D. inversely congruent.

E. None of the above.

11. Table Exam-2 portrays the truth values for a certain operation in propositional logic, denoted with a pound sign (#). Which of the following symbols belongs in place of the pound sign?

TABLE EXAM-2 Table for Final Exam Question 11.

X	Y	X # Y
F	F	F
F	T	F
T	F	F
T	T	T

A. An arrow pointing to the right (→).

B. A double-headed arrow (↔).

C. A wedge (∨).

D. An inverted wedge (∧).

E. None of the above.

12. Any given set *invariably* constitutes a subset of

A. any set with the same number of elements as itself.

B. any set with more elements than itself.

C. the null set.

D. itself.

E. None of the above.

13. Any given set *invariably* constitutes a *proper* subset of

A. any set with the same number of elements as itself.

B. any set with more elements than itself.

C. the null set.

D. itself.

E. None of the above.

14. A weak theorem typically takes the form of

A. a universal statement.

B. an existential statement.

C. an if/then statement.

D. a logical disjunction.

E. a logical injunction.

15. In pure mathematics and logic, proofs rely on inferences to establish significant final results called

A. conjunctions.

B. disjunctions.

C. theorems.

D. implications.

E. corollaries.

16. As we develop a new theory of numbers, we might use mathematical induction to

A. eliminate ridiculous notions about negative numbers.

B. prove a theorem about all negative integers.

C. describe a characteristic of all real numbers.

D. suggest an axiom concerning all natural numbers.

E. derive a contradiction to prove some assumption false.

17. Consider the boolean statement $(-X) \times (-Y)$. Which of the following statements is logically equivalent to it?

A. $-(Y + X)$

B. $-(Y \times X)$

C. $-(Y \Rightarrow X)$

D. $Y + (-X)$

E. $-(X \Rightarrow Y)$

TABLE EXAM-3 Table for Final Exam Question 18.

X	Y	X # Y
F	F	T
F	T	F
T	F	F
T	T	T

18. Table Exam-3 portrays the truth values for a certain operation in propositional logic, denoted with a pound sign (#). Which of the following symbols belongs in place of the pound sign?

A. An arrow pointing to the right (→).

B. A double-headed arrow (↔).

C. A wedge (∨).

D. An inverted wedge (∧).

E. None of the above.

19. Imagine that you're reading a paper and you come across the statement, "Let *x* be a variable in some universe *U*." This statement means that the author wants you to

A. consider the possibility that *x* might constitute a variable in *U*.

B. suppose, for the sake of argument, that *x* constitutes a variable in *U*.

C. acknowledge the fact that *x* might take some values outside of *U*.

D. believe that *x* can create or generate *U*.

E. understand that *x* is a subset of *U*.

20. The decimal quantity 64 equals the octal quantity

A. 64.

B. 32.

C. 77.

D. 100.

E. 111.

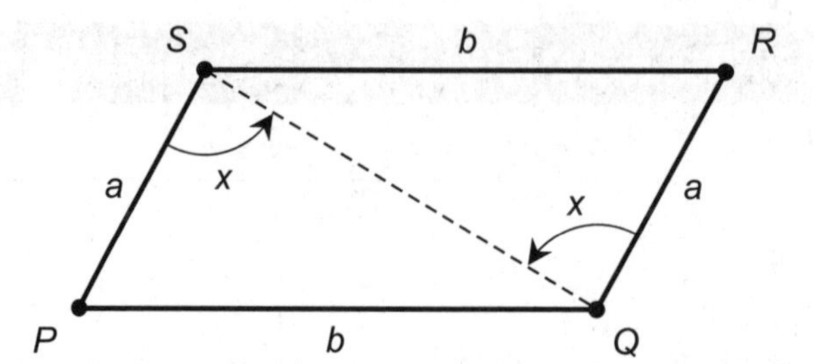

FIGURE EXAM-2 • Illustration for Final Exam Questions 21 and 22.

21. **Suppose that the figure *PQRS* in Fig. Exam-2 has sides and angles of lengths and measures as shown. What can you do if you want to prove that Δ*QSP* is directly congruent to Δ*SQR*?**

    A. You can prove it using the side-side-side (SSS) axiom.

    B. You can prove it using the side-angle-side (SAS) axiom.

    C. You can prove it using either the SSS axiom or the SAS axiom.

    D. You can prove it, but not using either the SSS axiom or the SAS axiom.

    E. You can't prove it, because it does not hold true in general.

22. **Suppose the figure *PQRS* in Fig. Exam-2 has sides and angles of lengths and measures as shown. What can you do if you want to prove that Δ*QSP* is inversely congruent to Δ*SQR*?**

    A. You can prove it using the side-side-side (SSS) axiom.

    B. You can prove it using the side-angle-side (SAS) axiom.

    C. You can prove it using either the SSS axiom or the SAS axiom.

    D. You can prove it, but not using either the SSS axiom or the SAS axiom.

    E. You can't prove it, because it does not hold true in general.

23. **Table Exam-4 portrays the truth values for a certain operation in propositional logic, denoted with a pound sign (#). Which of the following symbols belongs in place of the pound sign?**

    A. An arrow pointing to the right (→).

    B. A double-headed arrow (↔).

    C. A wedge (∨).

    D. An inverted wedge (∧).

    E. None of the above.

TABLE EXAM-4 Table for Final Exam Question 23.

X	Y	X # Y
F	F	F
F	T	T
T	F	T
T	T	F

24. Table Exam-5 shows a proof of the fact that in propositional logic, the order of a disjunction between two statements doesn't make any difference. In the "rule" column on the fifth line, a blank space appears. What should we write in this space to make the table complete?

 A. $\vee$E

 B. $\vee$I

 C. $\wedge$E

 D. $\wedge$I

 E. $\leftrightarrow$I

TABLE EXAM-5 Table for Final Exam Question 24.

Line number	Proposition	Reference	Rule	Assumption
1	P ∨ Q		A	1
2	P		A	2
3	Q ∨ P	2	∨I	2
4	Q		A	4
5	Q ∨ P	4		4
6	Q ∨ P	1, 2, 3, 4, 5	∨E	1

25. In Euclidean geometry, the parallel axiom tells us that

A. given a straight line and a point not on that line, there exists exactly one straight line that passes through the point and runs parallel to the original line.

B. given a straight line and a point not on that line, there exist no straight lines that pass through the point and run parallel to the original line.

C. given a straight line and a point not on that line, there exist exactly two straight lines that pass through the point and run parallel to the original line.

D. given a straight line and a point not on that line, there exist infinitely many straight lines that pass through the point and run parallel to the original line.

E. None of the above.

26. A compound sentence in propositional logic consisting entirely of statements joined by conjunctions holds true *if and only if*

A. at least one of its components is true.

B. more than half of its components are true.

C. all of its components are true.

D. all of its components are false.

E. at least one of its components is false.

27. We can define the boolean NAND operation as the

A. conjunction of two variables, followed by negation.

B. negation of a single variable, followed by conjunction.

C. negation of a single variable, followed by inclusive disjunction.

D. inclusive disjunction of two variables, followed by negation.

E. negation of a single variable, followed by exclusive disjunction.

28. Which, if any, of the following approaches A, B, C, or D, if any, will *not* suffice to prove direct congruence for triangles?

A. Side-side-side (SSS).

B. Angle-angle-angle (AAA).

C. Side-angle-side (SAS).

D. Angle-side-angle (ASA).

E. Any of the above approaches will suffice.

29. In the propositional calculus, which of the following binary connectives A, B, or C, if any, obeys the commutative law?

A. $\wedge$

B. $\vee$

C. $\leftrightarrow$

D. All of the above A, B, and C

E. None of the above A, B, or C

30. Which of the following statements A, B, or C, if any, fails to hold true in general for boolean variables X, Y, and Z?

A. $X + (Y \times Z) = (X + Y) \times Z$.

B. $X \times (Y \times Z) = (X \times Y) \times Z$.

C. $X + (Y + Z) = (X + Y) + Z$.

D. None of the statements A, B, or C hold true in general.

E. All of the statements A, B, and C hold true in general.

31. Consider two sets G and *H*, such that $G \cap H = \varnothing$. In this situation,

A. G and *H* overlap.

B. G and *H* are disjoint.

C. G and *H* share a single element.

D. G and *H* are congruent.

E. G and *H* are nondenumerable.

32. Logic works especially well for the purpose of assigning truth or falsity to

A. questions.

B. commands.

C. declarative sentences.

D. expressions of feeling.

E. All of the above.

33. If we want to "prove a negative," for example the proposition that no smallest integer exists, we should consider using the principle of

A. *modus ponens.*

B. *modus tollens.*

C. DeMorgan.

D. the contrapositive.

E. *reductio ad absurdum.*

34. Suppose that an astronomer tells you, "According to the most recent data we have, chances are better than 95% that at least one earth-like planet exists in our galaxy but outside our solar system." From a pure logician's point of view, this statement constitutes an example of

A. mathematical induction.

B. the law of implication reversal.

C. the probability fallacy.

D. *reductio ad absurdum.*

E. *modus ponens.*

35. Which of the following rules A, B, or C, if any, constitutes qualifications for the construction of a propositional formula (PF)?

A. Any propositional variable is a PF.

B. Any PF preceded by the symbol $\neg$ is a PF.

C. Any pair of PFs joined by one of the symbols $\wedge$, $\vee$, $\rightarrow$, or $\leftrightarrow$ is a PF.

D. All of the above rules A, B, and C constitute qualifications for the construction of a PF.

E. None of the above rules A, B, or C constitutes a qualification for the construction of a PF.

36. In Fig. Exam-3, which of the following statements holds true, assuming *A*, *B*, *C*, and *D* are all nonempty sets?

A. Sets *A* and *B* are disjoint, and sets C and *D* are disjoint.

B. Sets *A* and C are disjoint, and sets *B* and *D* are disjoint.

C. Sets *B* and C are disjoint, and sets *A* and *D* are disjoint.

D. Sets *A* and C are disjoint, and sets *A* and *D* are disjoint.

E. None of the above.

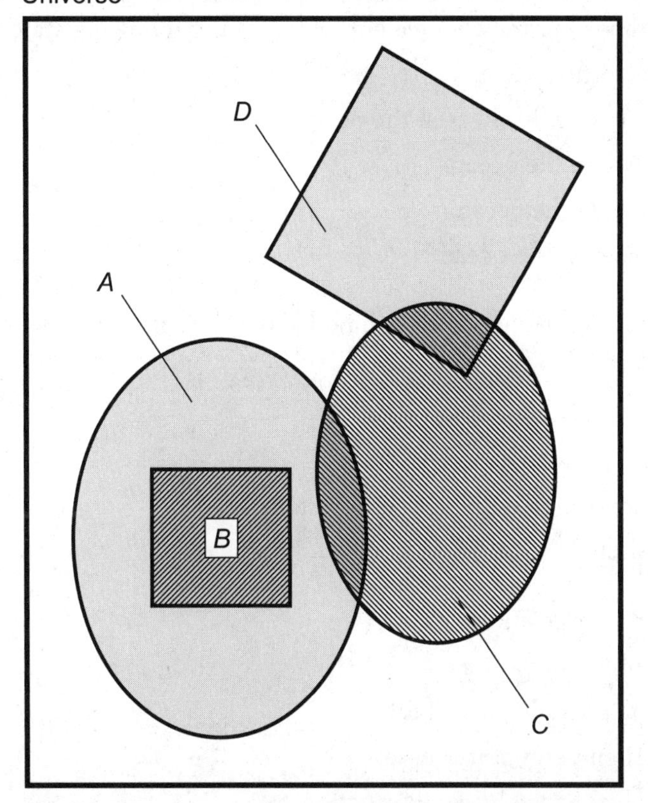

FIGURE EXAM-3 • Illustration for Final Exam Questions 36 and 37.

37. In Fig. Exam-3, which of the following statements holds true, assuming *A, B,* C, and *D* are all nonempty sets?

A. $B \subseteq A$

B. $B \in A$

C. $A \cap B = A$

D. $A \cap B = \varnothing$

E. $A \cup B = B$

38. Consider the proposition "There exists an object z such that z is a fish and z lives in the ocean." This proposition is logically equivalent to one of the following statements. Which one?

A. No fish live in the ocean.

B. One and only one fish lives in the ocean.

C. Some fish live in the ocean.

D. Most fish live in the ocean.

E. Every fish lives in the ocean.

39. In boolean algebra, we commonly symbolize the logical NOT operation as

A. a minus sign.

B. a backslash.

C. an addition sign with a circle around it.

D. a multiplication sign with a circle around it.

E. a forward slash.

40. An undecidable proposition

A. can exist only in fuzzy logic.

B. can never be proven true or false.

C. holds true if and only if it is false.

D. always involves an exclusive disjunction.

E. can be proven only by *reductio ad absurdum*.

41. Consider two sets G and *H*, such that $G \cap H = G$. In this situation, we can be *certain* that

A. G constitutes a subset of *H*.

B. G and *H* are disjoint.

C. G and *H* share a single element.

D. G and *H* are congruent.

E. *H* constitutes a subset of G.

42. The binary quantity 1100 equals the decimal quantity

A. 124.

B. 75.

C. 63.

D. 48.

E. 12.

43. What rule in propositional logic allows us to derive the proposition on either side of the symbol $\wedge$ by itself?

A. Conjunction introduction

B. Disjunction introduction

C. Conjunction elimination

D. Disjunction elimination

E. No existing rule

44. How can the elements in an infinite sequence of positive numbers add up to a finite positive number?

A. Such a thing can never happen, although we can trick ourselves into thinking that it sometimes can.

B. It can happen if the sequence follows the distributive law.

C. It can sometimes happen if the individual elements get smaller as we continue along in the sequence.

D. It can sometimes happen if the individual elements get larger as we continue along in the sequence.

E. It can happen if the numbers in the sequence approach infinity.

45. Which of the following constitutes an advantage of digital communications over analog communications?

A. We must copy an analog signal before its quality reaches an acceptable level, but we don't have to copy a digital signal at all.

B. Analog signals are restricted to binary form, while digital signals can attain infinitely many different states.

C. Digital errors occur less often than analog imperfections.

D. Digital signals are more sensitive to noise and interference than analog signals.

E. All of the above.

46. A compound sentence in propositional logic consisting entirely of statements joined by inclusive disjunctions holds true *if and only if*

A. at least one of its components is true.

B. more than half of its components are true.

C. all of its components are true.

D. all of its components are false.

E. at least one of its components is false.

47. A megabyte equals

A. 1000 bytes.

B. 1024 bytes.

C. 65,536 bytes.

D. 1,000,000 bytes.

E. 1,048,576 bytes.

48. One of the most famous axioms in geometry states that if you have a line *L* and a point *P* not on line *L*, then there exists one and only one line *M* through point *P* that is parallel to line *L*. But some mathematicians investigated the consequences of denying this axiom. It turns out that the axiom does not hold true

A. on flat surfaces.

B. when all lines are perfectly straight.

C. on the surface of a sphere.

D. in Euclidean three-dimensional space.

E. in any case; it was a huge mistake to begin with.

49. Consider the following proof, expressed in everyday language using three distinct statements on three lines.

All fish live in water.

All trout are fish.

Therefore, all trout live in water.

This argument constitutes an example of

A. a categorical syllogism.

B. an existential derivation.

C. a universal derivation.

D. a double implication.

E. an equivalence relation.

50. In the formal system of Euclidean geometry, what do we call things that do not have any formal definition?

A. Tautological objects

B. Null objects

C. Elementary objects

D. Reflexive objects

E. Trivial objects

51. In the positive-logic circuit of Fig. Exam-4, the output state is high

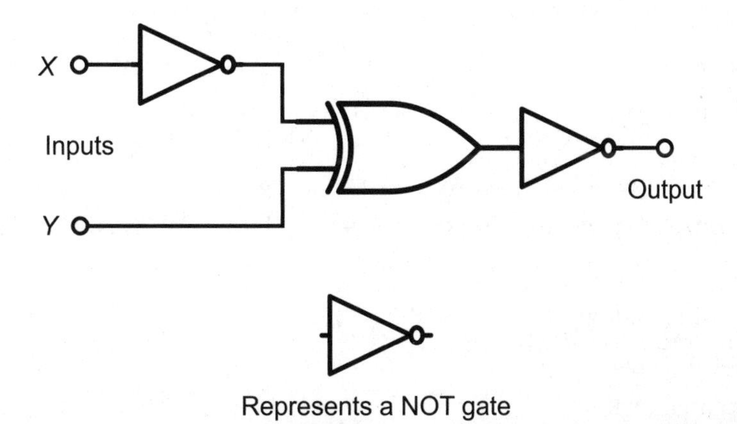

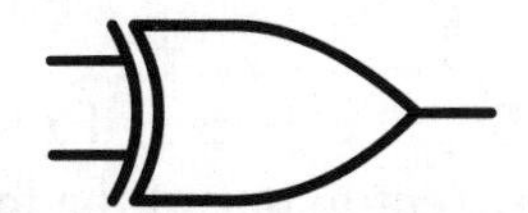

Represents an exclusive-OR (XOR) gate

FIGURE EXAM-4 • Illustration for Final Exam Questions 51 and 52.

A. when the input states are the same.

B. when input *X* is high and input *Y* is low.

C. when input *X* is low and input *Y* is high.

D. when either B or C holds true.

E. under no circumstances.

52. In the positive-logic circuit of Fig. Exam-4, the output state is low

A. when the input states are the same.

B. when input X is high and input Y is low.

C. when input X is low and input Y is high.

D. when either B or C holds true.

E. under no circumstances.

53. Boolean addition represents the equivalent of propositional-logic

A. disjunction.

B. negation.

C. implication.

D. conjunction.

E. equivalence.

54. A neutral truth state—that is, neither true nor false—can exist in a system of

A. trinary logic.

B. propositional logic.

C. predicate logic.

D. existential logic.

E. binary logic.

55. Consider the proposition "For all objects z, if z is a fish then z lives in the ocean." This proposition is logically equivalent to one of the following statements. Which one?

A. No fish live in the ocean.

B. A single fish lives in the ocean.

C. Some fish live in the ocean.

D. Most fish live in the ocean.

E. Every fish lives in the ocean.

56. Strong correlation between two events logically implies that

A. one causes the other.

B. a third event causes them both.

C. neither one can cause the other.

D. both events cause a third event.

E. None of the above.

57. When we see a turnstile symbol (⊦) between two sentences in propositional logic, we can read that symbol out loud as

A. "… if and only if …"

B. "… therefore …"

C. "…not …"

D. "… and …"

E. "… either-or …"

58. Which of the following sentences constitutes a statement of identity?

A. John drives to the store.

B. You and I share the same house.

C. A motorcar is an automobile.

D. Hawaii is located in the tropics.

E. The wind blows at a full gale.

59. In predicate logic, we can't write out a quantifier's equivalent conjunction or disjunction completely when

A. the universe has a finite number of elements.

B. we give every element in a finite universe a proper name.

C. the universe is infinitely large.

D. the law of excluded middle applies.

E. the commutative laws of conjunction or disjunction apply.

60. Consider a relation H that holds true among three variables *x*, *y*, and *z* in the following manner:

$$(\forall x, y, z)\ Hxy \wedge Hyz \rightarrow Hxz$$

On this basis of this statement, we know that the relation H is

A. symmetric.

B. reflexive.

C. commutative.

D. transitive.

E. associative.

61. We can easily use a Venn diagram in predicate logic to

A. define an elementary object.

B. test the validity of a syllogism.

C. introduce an axiom.

D. demonstrate the law of double negation.

E. demonstrate DeMorgan's laws.

62. In the statement "Steven is warm, but Sherri is cold," the word "but" has the same logical meaning as one of the following symbols. Which one?

A. $\leftrightarrow$

B. $\vee$

C. $=$

D. $\wedge$

E. $\rightarrow$

63. What is the union of the sets $A = \{1, 2, 3\}$ and $B = \{3, 4, 5\}$?

A. The set $\{3\}$

B. The set $\{1, 2, 3\}$

C. The set $\{3, 4, 5\}$

D. The set $\{1, 2, 3, 4, 5\}$

E. The null set

64. What is the intersection of the sets $A = \{1, 2, 3\}$ and $B = \{3, 4, 5\}$?

A. The set $\{3\}$

B. The set $\{1, 2, 3\}$

C. The set $\{3, 4, 5\}$

D. The set $\{1, 2, 3, 4, 5\}$

E. The null set

65. What rule in the sentential calculus lets us derive a $\wedge$-statement from a pair of propositions?

A. Conjunction introduction

B. Disjunction introduction

C. Conjunction elimination

D. Disjunction elimination

E. No existing rule

66. **Boolean implication represents the equivalent of propositional-logic**
   A. disjunction.
   B. negation.
   C. implication.
   D. conjunction.
   E. equivalence.

67. **Imagine that you invent a new mathematical relation among geometric objects. You call this relation "yotto-congruence" and symbolize it with a Japanese yen symbol (¥). Suppose you manage to prove that for any three geometric figures Q, *R*, and *S*, the following three properties hold true:**

$$Q \yen Q$$

$$(Q \yen R) \rightarrow (R \yen Q)$$

$$[(Q \yen R) \wedge (R \yen S)] \rightarrow (Q \yen S)$$

   From this, by definition, yotto-congruence constitutes
   A. a reflexive relation only.
   B. a transitive relation only.
   C. a commutative relation only.
   D. an existential relation.
   E. an equivalence relation.

68. **Which of the following four statements A, B, C, or D, if any, *does not* state a valid law of quantifier transformation where F represents a predicate and *x* represents a variable?**
   A. $(\exists x)\ Fx \dashv\vdash \neg(\forall x)\ \neg Fx$.
   B. $(\forall x)\ Fx \dashv\vdash \neg(\exists x)\ \neg Fx$.
   C. $\neg(\exists x)\ Fx \dashv\vdash (\forall x)\ \neg Fx$.
   D. $\neg(\forall x)\ Fx \dashv\vdash (\exists x)\ \neg Fx$.
   E. All four of the above statements A, B, C, and D state valid laws of quantifier transformation.

69. **How does inductive reasoning compare to mathematical induction?**
   A. Inductive reasoning constitutes an accepted rigorous method of proof, but mathematical induction does not.
   B. Inductive reasoning involves the use of *modus ponens*, while mathematical induction takes advantage of *modus tollens*.
   C. Inductive reasoning relies on the distributive law of conjunction over disjunction, while mathematical induction relies on the distributive law of disjunction over conjunction.
   D. Mathematical induction constitutes an accepted rigorous method of proof, while inductive reasoning does not.
   E. They're identical; the terms refer to precisely the same rigorous technique for proving theorems.

70. **Consider two sets G and H, such that $G \cap H = G \cup H$. In this situation, we can be *certain* that**
   A. G constitutes a proper subset of H.
   B. G and H are disjoint.
   C. G and H share a single element.
   D. G and H are congruent.
   E. H constitutes a proper subset of G.

71. **We can define the boolean NOR operation as the**
   A. conjunction of two variables, followed by negation.
   B. negation of a single variable, followed by conjunction.
   C. negation of a single variable, followed by inclusive disjunction.
   D. inclusive disjunction of two variables, followed by negation.
   E. negation of a single variable, followed by exclusive disjunction.

72. **Suppose that someone delivers the following argument: "My dog is green, and all green dogs were born on Mars. Therefore, my dog was born on Mars." What purely logical flaw, if any, exists in this argument?**
   A. The premises are ridiculous.
   B. The conclusion is ridiculous.
   C. The premises do not logically imply the conclusion.
   D. The conclusion contradicts the premises.
   E. No purely logical flaw exists in this argument.

73. The hexadecimal quantity B16 equals the decimal quantity

A. 2838.

B. 1967.

C. 897.

D. 257.

E. 129.

74. Your friend claims that "Some edible vegetables are not plants." You doubt the truth of this statement, and you want to rigorously demonstrate its falsity. In order to do that, you must prove that

A. some plants are not edible vegetables.

B. all plants are edible vegetables.

C. all edible vegetables are plants.

D. no edible vegetables are plants.

E. no plants are edible vegetables.

75. In the propositional calculus, which of the following binary connectives A, B, or C, if any, obey the associative law?

A. $\wedge$

B. $\vee$

C. $\rightarrow$

D. More than one of the above A, B, and C

E. None of the above A, B, or C

76. When we see a turnstile symbol ($\vdash$) at the extreme left-hand end of a statement in propositional logic, we can read that symbol out loud as

A. "Let's assume that ..."

B. "It is not true that ..."

C. "It is sometimes true that ..."

D. "It is a theorem that ..."

E. nothing, because placing the turnstile at the extreme left-hand end of a statement constitutes an improper use of that symbol.

77. A terabit equals

A. 1000 bits.

B. 1,000,000 bits.

C. 1,000,000,000 bits.

D. 1,000,000,000,000 bits.

E. 1,000,000,000,000,000 bits.

78. Which of the following statements A, B, C, or D, if any, constitutes an accurate verbal description of the Pythagorean theorem?

A. Given a right triangle, the square of the length of the hypotenuse equals the sum of the squares of the lengths of the other two sides, if all lengths are expressed in the same units.

B. Given a right triangle, the length of the hypotenuse equals the sum of the lengths of the other two sides, if all lengths are expressed in the same units.

C. Given a right triangle, the length of the hypotenuse equals the sum of the squares of the lengths of the other two sides, if all lengths are expressed in the same units.

D. Given a right triangle, the square of the length of the hypotenuse equals the sum of the lengths of the other two sides, if all lengths are expressed in the same units.

E. None of the above statements A, B, C, or D provides us with an accurate verbal description of the Pythagorean theorem.

79. What does the device marked X in Fig. Exam-5 do?

A. It smooths out the imperfections in the input signal, producing an output signal of superior quality.

B. It receives simultaneous bits from multiple channels and retransmits them one at a time along a single line.

C. It receives bits from a single line and retransmits them in batches of simultaneous bits across multiple channels.

D. It receives parallel data bytes (octets) and converts them into an analog signal.

E. It receives serial data bytes (octets) and converts them into analog channels.

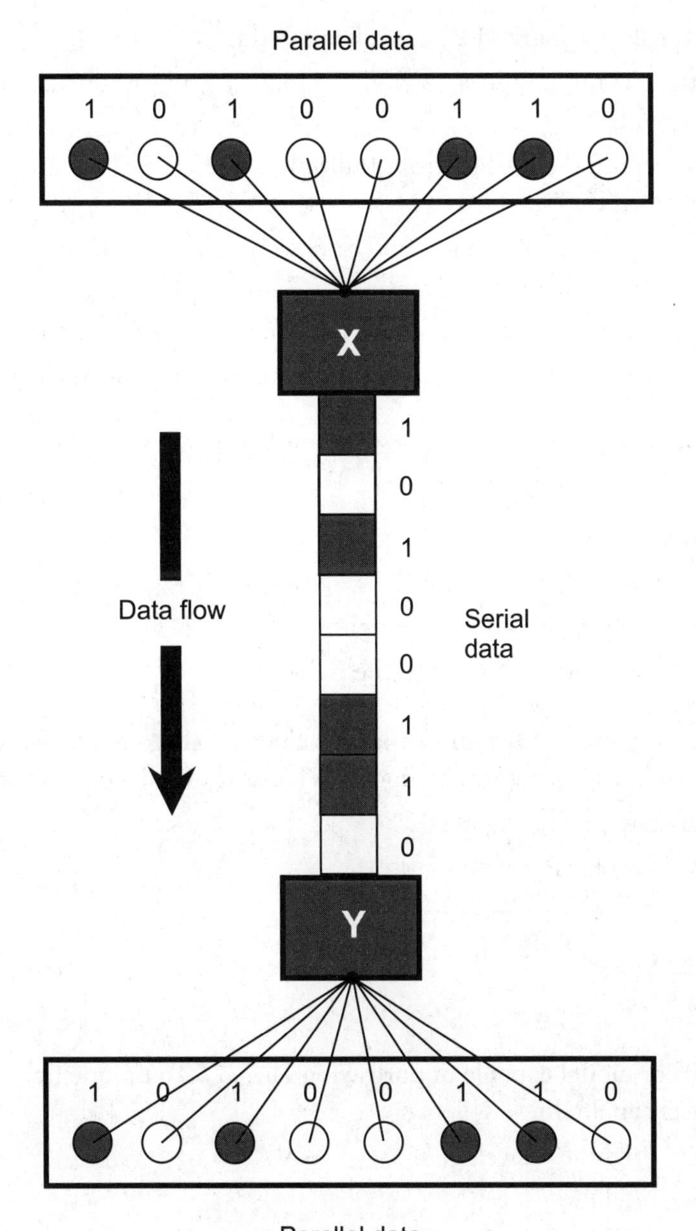

FIGURE EXAM-5 • Illustration for Final Exam Questions 79 and 80.

80. What does the device marked Y in Fig. Exam-5 do?

A. It smooths out the imperfections in the input signal, producing an output signal of superior quality.

B. It receives simultaneous bits from multiple channels and retransmits them one at a time along a single line.

C. It receives bits from a single line and retransmits them in batches of simultaneous bits across multiple channels.

D. It receives parallel data bytes (octets) and converts them into an analog signal.

E. It receives serial data bytes (octets) and converts them into analog channels.

81. In a graph, we usually plot the values of the independent variable

A. inside a circle of radius 1.

B. along the horizontal axis.

C. around a circle.

D. along a ray ramping up and to the right.

E. along a line ramping down and to the right.

82. Consider the argument "Obviously, that guy Stan Gibilisco never tells the truth. Look at him now, lying like a carpet on the floor!" What sort of fallacy does this claim commit?

A. Misuse of context

B. The probability fallacy

C. The fallacy of the contrapositive

D. DeMorgan's fallacy

E. Russell's fallacy

83. In an RGB color model capable of portraying 16,777,216 unique hues, each of the three color channels can attain

A. 16 discrete brightness levels.

B. 32 discrete brightness levels.

C. 64 discrete brightness levels.

D. 128 discrete brightness levels.

E. 256 discrete brightness levels.

84. The set of symbols and conventions that we can use to discuss the properties of propositional calculus constitutes

A. an irregular language.

B. a grammatical language.

C. a sequent language.

D. a well-formed language.

E. a metalanguage.

85. Consider two sets G and H, such that $G \cup H = \emptyset$. In this situation, we can be *certain* that

A. one of the sets contains infinitely many elements.

B. both of the sets are empty.

C. G and H share a single element.

D. G and H share more than one element.

E. one, but not necessarily both, of the sets are empty.

86. In fuzzy logic, truth can attain values

A. of 0 or 1 only.

B. of -1 or $+1$ only.

C. of -1, 0, or $+1$.

D. that range over a continuum.

E. of positive infinity ($+\infty$) or negative infinity ($-\infty$).

87. Suppose that P, Q, and R represent the vertices of an equilateral triangle. We can prove that P, Q, and R

A. all coincide.

B. all lie along a single line.

C. lie at equal distances from each other.

D. do not all lie in the same plane.

E. define a unique line segment.

88. **When we cannot write down the elements of a set, even in the form of an "implied list," then we call that set**

A. superdenumerable.

B. quasidenumerable.

C. nondenumerable.

D. infinitely denumerable.

E. transcendentally denumerable.

89. **Consider the following dilemma and argument, which constitutes an example of a fallacy in a syllogism:**

- **Every human being must eat a cup of beans daily or develop high blood pressure. You eat a cup of beans daily. Therefore, you will not develop high blood pressure.**

What's the source of the fallacy here?

A. We have misapplied one of DeMorgan's laws.

B. We have misapplied the law of the contrapositive.

C. We have used inclusive conjunction in one case and exclusive conjunction in another case.

D. We have confused logical implication with logical equivalence.

E. We have confused the inclusive and exclusive forms of disjunction.

90. **After we've proven a mathematical theorem, a few short steps might prove another related theorem known as**

A. a consequent.

B. an antecedent.

C. a contrapositive.

D. a corollary.

E. an injunction.

91. **If we create too many axioms in a mathematical system, we unnecessarily increase the risk that we'll eventually**

A. prove too many theorems.

B. fail to understand our definitions.

C. derive a contradiction.

D. get a theorem that we can't understand.

E. find that our theory lacks propositions.

92. Suppose that we can transpose the variables or constants in a two-part predicate-logic relation without changing the truth value. In that case, we can define the relation as

A. reflexive.

B. symmetric.

C. transitive.

D. associative.

E. distributive.

93. The intersection of the null set with any other set always equals

A. that other set.

B. the null set.

C. the set containing 0.

D. the set containing the null set.

E. the universal set.

94. We get into a serious logical paradox when we try to imagine

A. a set that contains no elements at all.

B. the set of all sets that are not members of themselves.

C. the set of all sets that are members of themselves.

D. the set of all possible sets in the entire universe.

E. the set of all numbers that are not equal to themselves.

95. After we have proved a sequent between two propositions "in both directions," we can call those two propositions

A. conjunctive.

B. disjunctive.

C. associative.

D. commutative.

E. interderivable.

96. Consider the boolean statement Y ⇒ −X. Which of the following statements is logically equivalent to it?

A. X ⇒ −Y

B. −X ⇒ −Y

C. X ⇒ Y

D. X = Y

E. None of the above

97. Suppose we want to prove that on a flat surface, we'll never encounter a triangle with straight-line sides, and whose interior-angle measures add up to 200°. What technique might we consider in attacking this problem?

A. Mathematical induction

B. Proof by example

C. *Reductio ad absurdum*

D. Inductive reasoning

E. The probability fallacy

98. The law of excluded middle assures us that

A. a statement cannot act as a conjunction and a disjunction "at the same time."

B. an implication cannot operate in both directions "at the same time."

C. the double negation of a statement has the same truth value as the original statement.

D. every statement is logically equivalent to itself.

E. any given statement must be either true or false.

99. In boolean algebra, we represent logical equivalence as

A. multiplication.

B. subtraction.

C. addition.

D. division.

E. None of the above.

100. Consider the following proof, expressed in everyday language using three distinct statements on three lines.

It is daytime or it is nighttime.
It is not daytime.
It is nighttime.

This argument constitutes an example of

A. a universal syllogism.

B. an existential syllogism.

C. a conjunctive syllogism.

D. a disjunctive syllogism.

E. a reflexive syllogism.

Answers to Quizzes and Final Exam

Chapter 1
1. D
2. B
3. C
4. A
5. D
6. A
7. A
8. C
9. B
10. D

Chapter 2
1. A
2. A
3. C
4. C
5. C
6. A
7. C
8. B
9. A
10. D

Chapter 3
1. A
2. C
3. C
4. B
5. B
6. C
7. D
8. A
9. D
10. C

Chapter 4
1. B
2. A
3. C
4. B
5. D
6. D
7. C
8. B
9. D
10. B

Chapter 5
1. C
2. A
3. A
4. A
5. B
6. B
7. C
8. D
9. D
10. C

Chapter 6
1. C
2. D
3. D
4. D
5. D
6. D
7. C
8. A
9. C
10. B

Chapter 7
1. A
2. C
3. D
4. D
5. D
6. B
7. B
8. D
9. C
10. B

Chapter 8
1. A
2. D
3. C
4. B
5. C
6. A
7. A
8. A
9. C
10. D

Chapter 9

1. D
2. A
3. B
4. B
5. C
6. A
7. D
8. D
9. A
10. C

Final Exam

1. E
2. D
3. C
4. D
5. A
6. B
7. A
8. E
9. C
10. A
11. D
12. D
13. E
14. B
15. C
16. B
17. A
18. B
19. B
20. D
21. C
22. E
23. E
24. B
25. A
26. C
27. A
28. B
29. D
30. A
31. B
32. C
33. E
34. C
35. D
36. C
37. A
38. C
39. A
40. B
41. A
42. E
43. C
44. C
45. C
46. A
47. E
48. C
49. A
50. C
51. D
52. A
53. A
54. A
55. E
56. E
57. B
58. C
59. C
60. D
61. B
62. D
63. D
64. A
65. A
66. C
67. E
68. E
69. D
70. D
71. D
72. E
73. A
74. C
75. D
76. D
77. D
78. A
79. B
80. C
81. B
82. A
83. E
84. E
85. B
86. D
87. C
88. C
89. E
90. D
91. C
92. B
93. B
94. B
95. E
96. A
97. C
98. E
99. E
100. D

Suggested Additional Reading

Carnap, R., *Introduction to Symbolic Logic and its Applications*. Dover Publications, New York, NY, 1958.

Cupillari, A., *The Nuts and Bolts of Proofs*, 2d ed. Academic Press, San Diego, CA, 2001.

Gibilisco, S., *Math Proofs Demystified*. McGraw-Hill, New York, NY, 2005.

Jacquette, D., *Symbolic Logic*. Wadsworth Publishing Company, Belmont, CA, 2001.

McNeill, D. and Freiberger, P., *Fuzzy Logic*. Simon & Schuster, New York, NY, 1993.

Mukaidono, M., *Fuzzy Logic for Beginners*. World Scientific Publishing Company, Singapore, 2001.

Priest, G., *Logic: A Very Short Introduction*. Oxford University Press, Oxford, England, 2000.

Sainsbury, R., *Paradoxes*, 2d ed. Cambridge University Press, Cambridge, England, 1992.

Solomon, A., *The Essentials of Boolean Algebra*. Research and Education Association, Piscataway, NJ, 1990.

Solow, D., *How to Read and Do Proofs*, 3d ed. John Wiley & Sons, Inc., New York, NY, 2002.

Velleman, D., *How to Prove It: A Structured Approach*. Cambridge University Press, Cambridge, England, 1994.

Whitesitt, J., *Boolean Algebra and Its Applications*. Dover Publications, New York, NY, 1995.

Zegarelli, M., *Logic for Dummies*. Wiley Publishing, Hoboken, NJ, 2007.

Index

A

affirmative proposition
  particular, 87–88
  universal, 87
alternate interior angles, 130
amplitude of signal, 265
analog signal, definition of, 265
analog vs. digital signal, 265
analog-to-digital conversion, 283–285
AND gate, 275–276
AND operation
  boolean, 216
  definition of, 9
angle
  definition of, 106–107
  included, 111
  measure of, 108
  right, 109
  straight, 108
  vertex of, 106–107
angle-side-angle axiom, 123–124
angles
  alternate interior, 130
  complementary, 109
  supplementary, 109
animated computer graphics, 284
antecedent
  boolean, 219
  confirming the, 41–42
  definition of, 11
  denying the, 157–158
antisymmetry, 76
arbitrary constant, 62
arbitrary name, 62
argument
  definition of, 1
  logical, 2–3
argument by circumstantial evidence, 151–152
arrow paradox, 161
ASCII, 277
associative laws, 50–52, 229
assumption, rule of, 37–38
asymmetry, 76
atom, 308–309
atomic sentence, 63
axioms
  in Euclidean geometry, 120–125
  logically consistent, 184
  in mathematical theory, 184

B

barbershop paradox, 173–174
base-2 number system, 267–268
base-8 number system, 268
base-10 number system, 266–267
base-16 number system, 268–269
baud, definition of, 282
Baudot, 277
begging the question, 150
biconditional
  associative law of, 51–52
  commutative law of, 49–50
biconditional elimination, 45
biconditional introduction, 44
big-bang theory, 311–312
binary connectives, 34

binary data, 277
binary number system, 267–268
binary system, 6
bit, definition of, 281–282
black box, 276, 278–281
boolean algebra, 215–238
boolean equation, definition of, 215
boolean expression, definition of, 215
brightness, 286
buffer, 286
butterfly effect, 323–324
byte, definition of, 282

C

categorical statement, 87
causation vs. correlation, 205–211
chaos theory, 319–330
circle, definition of, 182–183
circumstantial evidence, 151–152
classical paradoxes, 170–175
clock, 278
closed line segment, 100–101
closed-ended half-line, 102
closed-ended ray, 102
coincidence, 320–321
coincident lines, 103
collinear points, 103
color palette, 288–289
combinations of truth values, 19–20
commutative laws, 49–50, 228
compact disc, 284
complex logical operation, 18
compound logical operation, 18
complement, definition of, 7
complementary angles, 109
conclusion, definition of, 1
conditional proof, 42–43
confirming the antecedent, 41–42
congruence
  direct, 115–119
  inverse, 117–119
congruent sets
  definition of, 245–246
  intersection of, 253
  union of, 257
conjunction
  associative law of, 51–52
  boolean, 216–217
  commutative law of, 49–50
  definition of, 9
  DeMorgan's law of, 53
  elimination, 39
  grouping of, 21–22
conjunction (*Cont.*):
  introduction, 39
  reversing the order of, 21
  of several sentences, 13
  truth table for, 14, 221
  ungrouping the negation of, 25–26
consequent
  boolean, 219
  definition of, 11
  denying the, 42
constant, arbitrary, 62
context, misuse of, 151
continuum of truth, 147
contradiction
  law of, 48, 228
  principle of, 6
  in propositional logic, 35–36
contrapositive, law of, 229
contraries, 73–74, 92
convergent series, 163
conversion laws for syllogisms, 89
coordinates, 312
coplanar points, 103
copula, definition of, 7
corollary, definition of, 189–190
correlation
  vs. causation, 205–211
  vs. chaos, 320–321
  vs. coincidence, 320–321
  positive, 205–206
counter, 278

D

damped oscillation, 326–327
data speed comparisons, 283
decimal number system, 266–267
declarative sentence, definition of, 2
deduction, 2, 194–195
definitions in a mathematical theory, 182–183
DeMorgan's laws, 53, 230
denumerably infinite set, 241–242
denying the antecedent, 158–159
denying the consequent, 42
destination, 285
digital circuits, 274–281
digital signal
  bits in, 281–282
  definition of, 265
  states of, 290–291
digital signal processing, 286, 288–289
digital versatile disc, 284
digital vs. analog signal, 265
digital-to-analog conversion, 285

direct congruence, 115–119
direct similarity, 112–113, 119
direct-contradiction paradox, 160
disjoint sets
  definition of, 247–248
  intersection of, 253–254
  union of, 257
disjunction
  associative law of, 51–52
  boolean, 217–218
  commutative law of, 49–50
  DeMorgan's law of, 53
  elimination, 40–41
  exclusive, definition of, 10
  grouping of, inclusive, 23–24
  inclusive, definition of, 9–10
  introduction, 40
  reversing the order of, inclusive, 23
  of several sentences, 13–14
  truth table for, 15, 221–222
  ungrouping the negation of, inclusive, 26–27
disjunctive syllogism, 93
distributive laws, 54–55, 230
domain of discourse, definition of, 69
Doppler effect, 300
double negation, law of, 38, 49, 228
dwindling-displacement effect, 317–319
dyadic predicate, 64

E

Einstein, Albert, 299
electrical potential, 274
electron, 309
element of a set, definition of, 182
elementary objects, 100
elementary sentence, 63
elementary terms, 100, 183–184
elements of a set, 240–241
empty set
  definition of, 182
  intersection with, 253
  union with, 253
end points of line segment, 100–101
equality, boolean, 219–220
equals sign for truth value, 17
equation, boolean, definition of, 215
equilateral triangle, 111, 119–120
equivalence
  boolean, 219–220
  definition of, 12–13
  relation, 78
  truth table for, 16–17, 222–223
Euclid of Alexandria, 99
Euclidean geometry, 99–144, 184–186
Euclid's fifth postulate, 186
Euclid's postulates, 184–186
excluded end point, 101
excluded middle, law of, 6, 48–49
exclusive logical disjunction
  definition of, 10
  truth table for, 15
exclusive OR gate, 276
exclusive OR operation, definition of, 10
execution paradox, 170–171
existential elimination, 79–80
existential import, 92–93
existential introduction, 79
existential quantifier, 67–68
existential statement, 68
existentially quantified proposition, 68
expression, boolean, definition of, 215
extension axiom, 120
extraction of the square root, 329

F

fallacies
  argument by circumstantial evidence, 151–152
  begging the question, 150
  hasty generalization, 151
  misuse of context, 151
  probability fallacy, 146–147, 156
  "proof by example," 148–150
  syllogism fallacies, 152–154
figure, 88–89
finite set, 241–242
flash drive, 285
formal logic, 33
formal predicate, 62
formal subject, 62–63
formal system, 33
frequency counter, 278
frog-and-wall paradox, 162–165
fuzzy logic, 8, 147
fuzzy truth, 147
fuzzy worlds, 146–147

G

Galilei, Galileo, 299
gaming, 285
gate
  in binary digital logic, 275–276
  in frequency counter, 278
gate time, 278
Gauss, Carl Friedrich, 186

Gaussian geometry, 186
general sentence, 68
generalization, hasty, 151
geometry
  Euclidean, 99–144, 184–186
  Gaussian, 186
  Riemannian, 186
geometry trick, 165–167
gigabit, definition of, 281
gigabyte, definition of, 282
gigahertz, definition of, 278
Gödel, Kurt, 175, 177–178
greatest lower bound, 322–323

H

half-line
  closed-ended, 102
  open-ended, 102
half-lines
  collinear, 103
  parallel, 104–106
  perpendicular, 109
half-open line segment, 101
hard drive, 285
Hardy, G. H., 194
hasty generalization, 151
hertz, definition of, 278
hexadecimal number system, 268–269
hexagon, regular, 135–137
high state, 274–275
high-definition television, 285
hue, 286
hyperspace, 312–313
hypospace, 313–314
hypothetical syllogism, 94

I

identity
  assertion of, 65
  principle of, 5
identity elimination, 82
identity introduction, 81
IF/THEN operation
  boolean, 218–219
  definition of, 11–12
implication
  boolean, 218–219
  definition of, 11–12
  reversal, law of, 51, 53
  reversing the order of, 25, 51, 53
  truth table for, 15–16, 222
included angle, 111
included end point, 101
included side, 111
inclusive logical disjunction
  definition of, 9–10
  grouping of, 23–24
  reversing the order of, 23
  truth table for, 15
  ungrouping the negation of, 26–27
inclusive OR operation
  boolean, 217–218
  definition of, 10
Incompleteness Theorem, 177–178
independent variable, 206
individual variable, definition of, 62–63
induction, mathematical, 150, 201–202, 204–205
inductive reasoning, 155–156
inference, definition of, 2
infimum, 322–323
infinite sequence, 162–163
infinite series, 162–163
infinite set, 241–242
infinity-space, 316–317
integers, nonnegative, 242
interderivability, 54–55, 72–73
interference, 277
interior angles of triangle, 110
intersection of sets, 253–256, 261–262
intratransitivity, 78
invalid reasoning, definition of, 2
inverse congruence, 117–118
inverse similarity, 113–115, 120
inverter, 275–276
irrational number, definition of, 199
irreflexivity, 77
isosceles triangle, 111

K

Kepler, Johannes, 299
kilobit, definition of, 281
kilobyte, definition of, 282
kilohertz, definition of, 278

L

lemma, definition of, 189
light-beam conundrum, 298–299
line, as elementary term, 100
line segment
  closed, 100–101
  end points of, 100–101
  half-open, 101

line segment (*Cont.*):
  length of, 102
  open, 101
line segments
  collinear, 103
  parallel, 104–105
  perpendicular, 109
lines
  coincident, 103
  parallel, 104
  perpendicular, 109
linking verb, definition of, 7
logic gates, 275–276
logical argument, 2–3
logical conjunction
  definition of, 9
  of several sentences, 13
  truth table for, 14
logical disjunction
  exclusive, definition of, 10
  inclusive, definition of, 9–10
  of several sentences, 13–14
  truth table for, 15
logical equivalence
  definition of, 12–13
  truth table for, 15–16, 222–223
logical form, 4–5
logical function, 276
logical implication
  definition of, 11–12
  truth table for, 15–16
logical inverter, 275–276
logical negation
  definition of, 8–9
  truth table for, 14
logical variable, definition of, 62–63
low state, 274–275
lower bound, 318, 322

M

machine logic, 265–296
major premise, 88–89
major term, 88–89
Malthus, Thomas, 325
Malthusian model, 325–328
Mandelbrot, Benoit, 319
mark vs. space, 277
mathematical induction, 150, 201–202, 204–205
mathematical theory
  definition of, 181
  definitions in, 182–183
  evolution of, 182–187, 193–194
measure of angle, 108
megabit, definition of, 281
megabyte, definition of, 282
megahertz, definition of, 278
memory, 278
metalanguage, definition of, 36
middle term, 88–89
minor premise, 88–89
minor term, 88–89
misuse of context, 151
mixed operations, regrouping with, 27–28
modus ponens, 41–42, 94
modus tollens, 42
mood, 89–90
Morse code, 277
multiple quantifier, 70–71
multiplication, boolean, 216–217
Murray code, 277

N

NAND gate, 275–276
negation
  boolean, 216
  definition of, 8–9
  truth table for, 14, 221
negative logic, 275
negative proposition
  particular, 88
  universal, 87
nesting of operations, 18, 227
neutrinos, 309
Newton, Isaac, 299
noise, 277
nondisjoint sets
  definition of, 248–250
  intersection of, 254–256
  union of, 258–259
nonnegative integers, 242
nonreflexivity, 77
nonsymmetry, 76
nontransitivity, 78
NOR gate, 275–276
NOT gate, 275–276
NOT operation
  boolean, 216
  definition of, 8–9
nucleus of atom, 308–309
null set
  definition of, 182
  intersection with, 253
  union with, 253

number systems
  base-2, 267–268
  base-8, 268
  base-10, 266–267
  base-16, 268–269
  binary, 267–268
  decimal, 266–267
  hexadecimal, 268–269
  octal, 268
  radix-2, 267–268
  radix-8, 268
  radix-10, 266–267
  radix-16, 268–269

O

object, definition of, 7
octal number system, 268
one-to-one correspondence, 247
open line segment, 101
open-ended half-line, 102
open-ended ray, 102
operations, nesting of, 18
opposition square of, between quantified statements, 73–74
OR gate, 275–276
OR operation
  exclusive, 10, 218
  inclusive, 10, 217–218
order, ultimate state of, 311–312
order from randomness, 310–311
overlapping sets
  definition of, 248–250
  intersection of, 254–256
  union of, 258–259

P

paradoxes
  arrow paradox, 161
  barbershop paradox, 173–174
  classical paradoxes, 170–175
  direct contradiction, 160
  execution paradox, 170–171
  frog-and-wall paradox, 162–165
  geometry trick, 165–167
  "proof" that −1 = 1, 167–169
  Russell's paradox, 175–176
  saloon paradox, 172–173
  shark paradox, 174–175
  two-pronged defense paradox, 171–172
  wheel paradox, 169–170

paradoxes (*Cont.*):
  who-shaves-Hap paradox, 161
  wire around the earth, 158–160
parallel axiom, 121
parallel data transmission, 286–287
parallel half-lines, 104–106
parallel line segments, 104–105
parallel lines, 104
parallel postulate, Euclid's, 186
parallel rays, 104–106
parallel worlds, 146–147
parallel-to-serial converter, 286–287
particles without end, 308–310
particular affirmative proposition, 87–88
particular negative proposition, 88
particular sentence, 68
perpendicular line segments, 109
perpendicular lines, 109
perpendicular rays, 109
plane, as elementary term, 100
plane geometry, 99–144
point, as elementary term, 100
point of intersection, 102–103
points
  collinear, 103
  coplanar, 103
polyadic predicate, 64
population-vs.-time function, 326
positive logic, 274–275
potential, electrical, 274
precedence of operations
  in boolean algebra, 227–228
  in formal logic, 18–19
predicate
  definition of, 6
  dyadic, 64
  formal, 62
  polyadic, 64
predicate calculus, 61
predicate letters, 62
predicate logic, 61–97
predicate proofs, 78–86
predicate sentence formulas, 63–64
predicate term, 88–89
premise
  definition of, 1
  major, 88–89
  minor, 88–89
primordial fireball, 312
principle of contradiction, 6
principle of identity, 5
probability fallacy, 146–147, 156
"proof" by example, 148–150

"proof" that −1 = 1, 167–169
proofs
 in Euclidean geometry, 125–141
 predicate, 78–86
 strategies for, 181–213
proper subset, 245–246, 250–251
proposition
 definition of, 2
 in mathematical theory, 187
 particular affirmative, 87–88
 particular negative, 88
 singular, 63
 undecidable, 175–178
 universal affirmative, 87
 universal negative, 87
propositional calculus, 33
propositional expression, 34
propositional formulas, 34–35
propositional logic, 33–59
Proxima Centauri, 303
pseudorandom numbers, 330
Pythagorean theorem, 188–189

Q

Q.E.D., 18
quantified statements, 66–67
quantifier
 existential, 67–68
 multiple, 70–71
 translating between, 72–73
 universal, 68
queue, 286
Quod erat demonstradum, 18

R

r factor, 326–328
radix-2 number system, 267–268
radix-8 number system, 268
radix-10 number system, 266–267
radix-16 number system, 268–269
randomness
 order from, 310–311
 true, 329–330
rational number, definition of, 198
ray
 closed-ended, 102
 open-ended, 102
rays
 collinear, 103
 parallel, 104–106
 perpendicular, 109
real number, definition of, 198
reasoning
 inductive, 155–156
 rules for, 5–6
red/green/blue color model, 288–289
reductio ad absurdum, 43–44, 199–201
reference frame, 302
reflexivity, 77
regular hexagon, 135–137
relation
 definition of, 64
 two-part, 75–78
representative instance, 69
Riemann, Bernhard, 186
Riemannian geometry, 186
right angle, 109, 120
right angle axiom, 120
right triangle, 112, 188–189
rigor, definition of, 99
rules for reasoning, 5–6
Russell, Bertrand, 175–176
Russell's paradox, 175–176

S

Sagan, Carl, 303
saloon paradox, 172–173
sampling rate, 284
sampling resolution, 284
saturation, 286
scale parallels, 324–325
scale-recurrent patterns, 321–322
schematic diagram, 279
sentence
 atomic, 63
 elementary, 63
 in everyday language, 2
 forms, 6–7
sentential calculus, 33
sentential formula, 34
sentential logic, 33
sequence, infinite, 162–163
sequent, definition of, 36
serial data transmission, 285–287
serial-to-parallel converter, 286–287
series
 convergent, 163
 infinite, 162–163
set
 definition of, 175, 182
 denumerably infinite, 241–242
 elements of, 240–241
 finite, 241–242

set (*Cont.*):
infinite, 241–242
Russell's paradox and, 175–176
set theory, 239–264
sets
congruent, 246–247
disjoint, 247–248
intersection of, 253–256, 261–262
nondisjoint, 248–250
one-to-one correspondence between, 247
overlapping, 248–250
Venn diagram and, 244–253
sets within sets, 242–243
shark paradox, 174–175
side, included, 111
side-angle-angle axiom, 123–125
side-angle-side axiom, 122–123
side-side-side axiom, 121
sides of triangle, 110
signal amplitude, 265
similarity
direct, 112–113, 119
inverse, 113–115, 119–120
simultaneity, 299–300
singular proposition, 63
singularity, 304
sound reasoning, definition of, 2
source, 285
space vs. mark, 277
sphere, definition of, 182–183
Spiral of Archimedes, 324–325
square of opposition
between quantified statements, 73–74
between syllogism types, 92
square root, extraction of, 329
steady-state theory, 311–312
straight angle, 108
strong theorem, 197–199
subcontraries, 73–74, 92
subject
definition of, 6
formal, 62–63
subject letters, 62
subject/linking verb/complement (SLVC) statement, 7
subject term, 88–89
subject/verb (SV) statement, 6–7
subject/verb/object (SVO) statement, 7
subset, 245–246, 250–252
substitution, 45–56
supplementary angles, 109
syllogism, 86–94
conversion laws for, 89
disjunctive, 93
syllogism (*Cont.*):
figure of, 89–90
hypothetical, 94
mood of, 89–90
terms in, 88–89
testing of, 90–93
validity of, 83
Venn diagram and, 90–91
syllogisms, fallacies with, 152–154
symbolic logic, 34–35
symbols for logical operations, 8–14
symmetry, 75–76

T

tachyon, 309
tautology, 150
terabit, definition of, 281
terabyte, definition of, 282
terahertz, definition of, 278
terms in a syllogism, 88–89
theorem
definition of, 3
in mathematical theory, 187–194
Pythagorean, 188–189
strong, 197–199
weak, 196–197
theorem introduction, 54–56
time, illogic of, 298–308
time as a dimension, 314–316
time dilation, 300–303
time travel, 303–304
time-space, 314
transitivity, 77
transversal, definition of, 104
triangle
definition of, 110
equilateral, 111
interior angles of, 110
isosceles, 111
right, 112
sides of, 110
vertices of, 110
trinary logic, 8
truth continuum, 147
truth table
boolean, 220–227
for conjunction, 14, 22
definition of, 14
for exclusive disjunction, 15
for implication, 15–16, 222
for inclusive disjunction, 15, 221–222
for logical equivalence, 15–16, 222–223

truth table (*Cont.*):
 for negation, 14, 221
 proofs using, 20–30
truth values
 combinations of, 18–19
 equals sign for, 17
twin paradox, 304–308
two-part relation, 75–78
two-point axiom, 120
two-pronged defense paradox, 171–172
two-state system, 6
typical conjunct, 69
typical disjunct, 69

U

undecidable proposition, 175–178
undefined terms, 183–184
union of sets, 256–262
universal affirmative proposition, 87
universal elimination, 81
universal introduction, 80–81
universal negative proposition, 87
universal quantifier, 68
universal set, 195
universal statement, 68
universally quantified proposition, 68
universe, definition of, 68–69, 195
unsound reasoning, definition of, 2

V

valid reasoning, definition of, 2
variable
 independent, 206
 logical, definition of, 62–63
Venn diagram
 sets and, 244–253, 260–261
 syllogism and, 90–91
vertex of angle, 106–107
vertices of triangle, 110
virtual reality, 284–285

W

weak theorem, 196–197
well-formed formulas, 74–75
wheel paradox, 169–170
whole numbers, 242
who-shaves-Hap paradox, 161
wire-around-the-earth paradox, 158–160

XYZ

XOR gate, 276
XOR operation, boolean, 218